GENETIC AND REPRODUCTIVE ENGINEERING

Edited by
Darrel S. English
Northern Arizona University

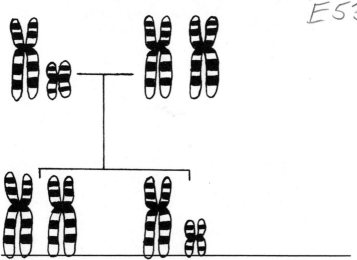

MSS Information Corporation
655 Madison Avenue, New York, N.Y. 10021

This is a custom-made book of readings prepared for the courses taught by the editor, as well as for related courses and for college and university libraries. For information about our program, please write to:

MSS INFORMATION CORPORATION
655 Madison Avenue
New York, New York 10021

MSS wishes to express its appreciation to the authors of the articles in this collection for their cooperation in making their work available in this format.

Library of Congress Cataloging in Publication Data

English, Darrel S   comp.
    Genetic and reproductive engineering.

    CONTENTS: Sutton, H.E. Human genetics: a survey of new developments. — Heller, J.H. Human chromosome abnormalities as related to physical & mental dysfunction. [etc.]
    1. Genetic engineering — Addresses, essays, lectures.
2. Reproduction — Addresses, essays, lectures.
I. Title. [DNLM: 1. Genetic intervention — Collected works. 2. Reproduction — Collected works. QH431 E58g 1974]
QH442.E53   616'.042   73-22048
ISBN 0-8422-5157-X
ISBN 0-8422-0383-4 (pbk.)

CONTENTS

BIOETHICS: WHO DECIDES WHAT

PREFACE

Unraveling the complexities of the double helix has opened the doors to the secrets of the genetic code, the process of protein synthesis and gene regulation. This information has been the basis upon which man has gained insight into the genetics of cancer, metabolic diseases, human development and even the depths of the mind. One may ask then, "What are the recent developments in genetics and where will they take us?"

The introductory four articles by Sutton, Heller, Libassi and Rucknagel cover the present art of detection and treatment of genetic abnormalities. Significant advancements have been made in the analysis of human chromosome numbers and their aberrations. Likewise, understanding the biochemical cause of deleterious mutations has resulted in successful treatments. More recently, new techniques such as chromosome banding, the use of ultrasonics, and computers, have teamed up to achieve greater advances in the elimination of genetic diseases. Although these developments enable medical science to locate the problem, the treatment of such defects is only temporary from a genetic viewpoint since the basic problem has not been eliminated and will be handed on to the next generation.

In the next three articles by English, Friedmann and Roblin, and Rabovsky, we consider some of the revolutionary possibilities of dealing with genetic disorders. Even though the techniques for altering one's genetic make-up in a specific direction has not been perfected, it appears that it is only a matter of time until such a goal will be realized. Actual "gene surgery" or genetic engineering holds the potential of not only alleviating the disease but permanently correcting the the defect for generations to come. This would be the ultimate in medical genetics. These techniques also make possible a far greater understanding of the workings of the basic machinery of the cell. Such research is not without its potential hazards, however, just as today's techniques and procedures also have their dangers. Some of these social, legal and ethical considerations are presented in the last section.

Three papers by Rivers, Beatty and Francoeur introduce the fourth section, which deals with a closely related area called reproductive engineering. Man may forget how subtly he has been involved in manipulating his own reproductive behavior, but he need only realize how quickly he has accepted as fact such procedures as artificial insemination, birth control, sperm banks, amniocentesis, genetic coun-

5

seling and to an increasing extent abortion, to realize his deep involvement. In the future such dramatic advances as inovulation, cloning and legislative control over human procreation may also be commonly accepted. Tremendous potential as well as liabilities are sure to be the legacy of such research in the future of mankind. This section closes with an article by Glass which considers the influences of genetic engineering on the human gene pool as it relates specifically to man's ability to survive and evolve in his environment.

The final section briefly deals with bioethics, that is, the promise and perils of the genetics of tomorrow. There is a growing concern that emerging techniques, enabling man to manipulate his genetic and evolutionary future, have outstripped his ability to deal with such procedures on a social, legal and ethical level. Will science and government use these techniques to exploit the unsuspecting public? Who will decide what group or agency will be endowed with the power to genetically determine our evolutionary future? Is there a danger of arousing public hopes and fears unnecessarily? Are these questions really relevant at this stage of progress in genetic engineering? Society was generally unprepared to handle the barrage of problems suddenly accompanying its entry into the Atomic Age. It is hoped that in the next several decades we will be more ready to deal with the Genetic Age.

This book is intended for both science and non-science majors. It is the conviction of the editor that people in all walks of life will sooner or later be facing some of these problems and it is hoped they will have had the opportunity to rationally consider some of the implications these new developments will have on future generations. This book is not intended to be an exhaustive coverage of the field. Many equally interesting articles have been omitted. Those articles chosen were considered to be readable, informative and thought provoking. The present state of genetic and reproductive engineering is certainly controversial and the author has not intentionally tried to convey any particular point of view.

Human Genetics:
A Survey of New Developments

H. ELDON SUTTON

IT IS almost exactly ten years ago that human cyto-genetics became a distinct discipline with emphasis on the relationship of chromosome constitution to human variation. Prior to that, efforts had been devoted primarily to development of techniques for chromosome study. By 1957, several procedures had become available, enabling direct examination and characterization of somatic chromosomes of most tissues. Foremost among these procedures were (1) the ability to culture cells *in vitro* at least for a short period (several cell divisions); (2) the use of colchicine to arrest cell division in metaphase, leading to the accumulation of many cells at the optimum stage for study; and (3) treatment of cells with buffers of low osmotic strength, resulting in dispersion of the chromosomes with much less overlapping.

Attention was dramatically focused on the new field of human cytogenetics by the report that human beings have only 46 chromosomes rather than the 48 that had been the accepted number for almost 35 years. These 46 consist of 22 homologous pairs, the autosomes, plus a pair of sex chromosomes (two X chromosomes in females and an X and Y chromosome in males). Chromosomes differ in length and in position of the centromere (or kinetochore), the point at which spindle fibers attach in division and which play a major role in assuring correct distribution of daughter chromosomes to new daughter cells. In somatic cell division (mitosis), colchicine halts chromosome division after the new chromosome arms have separated but before the centromere has divided. Typically, chromosomes prepared in this manner appear as variations on the letter X, the centromere being the point at which the four

THE SCIENCE TEACHER, 1967, Vol. 55, No. 15, pp. 51-55.

arms meet. (Figure 1) Its position may be anywhere along the chromosome, although in man there are no chromosomes with centromeres on the end. By the size of the chromosomes and position of the centromere, human chromosomes can be classified into seven groups, usually designated by letters A through G. Some specific chromosome pairs can be recognized consistently within these groups; others cannot. With rare exception, the chromosome complements (karyotypes) of normal persons are indistinguishable except for the XX or XY combination.

The announcement in 1959 that persons with mongolian idiocy have 47 rather than 46 chromosomes stimulated an enormous effort to examine chromosomes of many other disorders of obscure etiology. In mongolism (also called Down's syndrome), there are three rather than two of one of the G group chromosomes, often designated chromosome 21. This condition is therefore said to result from *trisomy* of a G chromosome and is called the trisomy 21 syndrome by some. The features of the condition, which include mental deficiency, a typical facial appearance, and characteristic palm prints, can be attributed to the unbalancing effect of three copies of many genes as opposed to the normal two copies.

Only two other autosomal trisomy syndromes are known, one involving a group D (13-15) chromosome, the other involving a group E (No. 18). Both are asso-

Figure 1. *Chromosome spread and karyotype of a normal human male.*

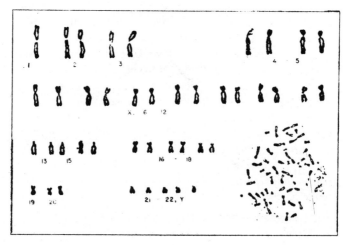

8

ciated with multiple malformations, and most affected infants do not survive more than a few weeks.

Abnormal numbers of sex chromosomes occur approximately once in 500 births. The most common is the combination XXY, which results in a male with Klinefelter's syndrome. Such persons may be essentially normal in appearance, although most often they have features which are somewhat eunochoid, and they are infertile. The most common defect in females is Turner's syndrome, or gonadal dysgenesis, rarer than Klinefelter's syndrome, with only 45 chromosomes including one X and no Y chromosome. The karyotype of such a person is written 44A + XO, the letter O indicating absence of the homologous chromosome.

The recognition that in human beings the XXY complement produces males and the XO females was an indication that sex is determined by presence or absence of a Y chromosome. This is in contrast to *Drosophila,* where the ratio of X chromosomes to autosomes determines the sex. Confirmation came with the discovery of cases with greater numbers of X chromosomes—XXXY and XXXXY. These also are males, very little different from XXY or, for that matter, XY males. The XYY karyotype has been observed a number of times and often results in increased height and aggressive behavior; otherwise, the person is essentially normal.

ACCIDENTS occasionally give rise to chromosomes of abnormal structure. Of particular importance are deletions and translocations. Deletions are losses of a chromosome segment and often arise when chromosomes break and fail to rejoin. Only those parts attached to a centromere are properly distributed in cell division, and segments distal to a chromosomal break are lost. The result is partial monosomy for the lost segment, with consequent genic imbalance and expression of recessive genes on the corresponding segment of the normal homologous chromosome.

Most major deletions are incompatible with life. Several have occurred a number of times, of which the *cri-du-chat* or "cat's cry" syndrome is the best known. This results from deletion of part of the short arm of chromosome 5, and affected persons are severely retarded mentally, with a characteristic facial appearance and often a peculiar cat-like wail when young. Two other deletions seen in a num-

9

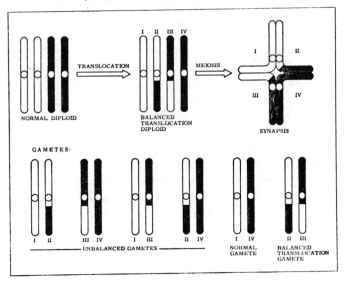

Figure 2. Translocation between two nonhomologous chromosomes showing pairing in meiosis and the six kinds of gametes theoretically possible. Two pairs of chromosomes are shown. Other chromosomes in the complement would behave normally. Crossing over in the translocated segment would produce additional combinations. Although six possibilities are shown, these need not be equally likely to occur.

ber of patients involve partial deletion of the short arm of chromosome 18 and partial deletion of its long arm.

One form of malignancy, chronic granulocytic leukemia, is due to partial deletion of the long arm of a G group chromosome, thought to be the same that is trisomic in mongolism. Such patients start out with a normal complement, but a deletion occurs in the marrow cells, giving rise to a clone of abnormal cells responsible for this form of leukemia. Other forms of malignancy, including the other leukemias, have not been associated with chromosomal abnormalities.

If breaks occur in two nonhomologous chromosomes, exchange of distal segments sometimes occurs during the repair process, and translocation is said to have occurred. The cell in which this happens still has a normal gene complement, but the genes are now associated into abnormal combinations. The cell and those that arise from it by simple mitotic division usually are normal, but the reduction division of meiosis can no longer occur normally. An illustration of what may happen is given in Figure 2.

Gametes from such a meiotic division may be normal, a balanced translocation, or an unbalanced combination. The balanced translocation, combined with a normal gamete, produces a zygote similar in constitution to the cell in which the translocation arose. The unbalanced combinations give rise to various types of partial monosomy and partial trisomy.

Translocations are of medical significance because they are a mechanism by which an apparently normal person (with a balanced translocation) produces a high frequency of abnormal offspring. These offspring may be similar to recognized chromosomal syndromes, or they may be classified only as miscarriages, being incompatible with development to live birth. Offspring who inherit the balanced translocation will be normal but will also produce abnormal offspring with high frequency. A number of pedigrees are known in which mongolism due to translocation imbalance has appeared repeatedly. The risk of defect is very much greater in offspring of balanced translocation carriers compared to normal persons, and it is important in estimating risk for future offspring to ascertain whether a child with mongolism, for example, has a simple trisomy with 47 chromosomes or an unbalanced translocation.

NOWHERE has human genetics made a greater contribution to general genetic understanding than in the area of biochemistry and biochemical genetics. Much of this understanding comes from studies of inherited variants of hemoglobin. Hemoglobin, like other proteins, consists in part of long chains of amino acids attached by peptide bonds. There are 20 amino acids commonly found in proteins, and the specific sequence determines the properties of the molecule. A complete hemoglobin molecule has four such polypeptide chains. Two of the chains are of one type, designated α chains; the other two are of a second type, designated β chains. There are 141 amino acids in each α chain and 146 in each β chain.

Some 60 genetic variants of human hemoglobin are known, most of them differing from the usual form because of a slight difference in charge and hence in electrophoretic mobility. The nature of the structural variations associated with different genes was established by Ingram in 1956 working with sickle cell hemoglobin (Hb S). This form of hemoglobin leads to serious disease if the person has two genes for it. Ingram showed that the

Table 1. Amino acid substitutions in inherited human hemoglobin variants. A number of additional variants have been studied but are not given in the table.

Amino acid residue No.	Residue in Hb A	Variant	
		Substitution	Designation of Hb
α chain			
1	val [a]		
2	leu		
16	lys	glu	I
30	glu	gln	G Honolulu
57	gly	asp	Norfolk
58	his	tyr	M Boston
68	asn	lys	G Philadelphia
87	his	tyr	M Iwate
116	glu	lys	O Indonesia
141	arg		
β chain			
1	val		
2	his		
3	leu		
6	glu	{ val	S
		{ lys	C
7	glu	gly	G San Jose
26	glu	lys	E
63	his	tyr	M Saskatoon
79	asp	asn	G Accra
121	glu	lys	O Arabia
146	his		

[a] Abbreviations of amino acids are as follows: arg (arginine), asn (asparagine), asp (aspartic acid), gln (glutamine), glu (glutamic acid), gly (glycine), his (histidine), leu (leucine), lys (lysine), tyr (tyrosine), and val (valine).

only difference between Hb S and the normal Hb A is in the amino acid in the sixth position of the β chain. In Hb A it is glutamic acid; in Hb S it is valine. Studies on some three dozen of the other variants have confirmed that each varies in a single amino acid at some specific point along the chain. (Table 1) From these studies, it is clear that one important action of genes is to determine the sequence of amino acids in polypeptide chains. Different forms of a gene produce different but closely related sequences, ordinarily differing only at a single amino acid position.

The mechanisms by which genes direct protein structure have been worked out largely on other organisms, but hemoglobin variants follow the predicted pattern. Genetic information is stored in the sequence of nucleotides in deoxyribonucleic acid (DNA). A sequence of three nucleotides constitutes a *codon*, i.e., the coding unit for one amino acid. Any of four nucleotides is possible at a given position in the DNA; hence there are 64 possible nucleotide sequences in a nucleotide triplet. This is more than the

Table 2. RNA code for amino acids. The RNA code is that which functions for messenger RNA and is complementary to the DNA code. Several lines of evidence have been used to establish the code, which appears to be universal biologically. Abbreviations for acids: U (uridylic), C (cytidylic), A (adenylic), G (guanylic).

Phenylalanine	Serine	Tyrosine	Glutamic acid
UUU	UCU	UAU	GAA
UUC	UCC	UAC	GAG
	UCA		
Leucine	UCG	Blank [a]	Cysteine
UUA	AGU	UAA	UGU
UUG	AGC	UAG	UGC
CUU			
CUC	Proline	Histidine	Tryptophan
CUA	CCU	CAU	UGA(?)
CUG	CCC	CAC	UGG
	CCA		
Isoleucine	CCG	Glutamine	Arginine
AUU		CAA	CGU
AUC	Threonine	CAG	CGC
AUA	ACU		CGA
	ACC	Asparagine	CGG
Methionine	ACA	AAU	AGA
AUG	ACG	AAC	AGG
Valine	Alanine	Lysine	Glycine
GUU	GCU	AAA	GGU
GUC	GCC	AAG	GGC
GUA	GCA		GGA
GUG	GCG	Aspartic acid	GGG
		GAU	
		GAC	

[a] These two codons are blank in the sense that growth of the polypeptide chain stops at the point at which they occur and the polypeptide chain is released as "completed." They may function therefore as punctuation.

20 amino acids incorporated into proteins, and it is now established that several codons may code for the same amino acid. (Table 2) The 141 amino acids of the hemoglobin α chain would correspond to a sequence of 141 nucleotide codons, which in turn would be $3 \times 141 = 423$ nucleotides.

Most mutations involve single codon changes. By comparison of the codons for glutamic acid against those of valine, it is seen that a single nucleotide exchange could give rise to the hemoglobin S mutation, e.g., GAA→GUA. Similar results are obtained with other hemoglobin variants. By contrast, many theoretically possible amino acid substitutions could arise only by changing two of the three codon nucleotides. Such variants are rare. These results confirm that the genetic code, worked out principally with microorganisms, is the same for man and presumably for all biological systems.

Some types of changes in the DNA result in interruption of the amino acid sequence so that a complete polypeptide chain cannot be formed. Rather than a slightly altered protein product, no recognizable product is formed. Such mutants, when homozygous, lack the function associated with the normal protein whose synthesis is directed by that gene locus.

Several types of changes in the DNA may give rise to loss of recognizable protein product. If a segment of DNA is deleted, a complete polypeptide chain cannot be formed, although part of the chain corresponding to the remaining nucleotides might be formed. Ordinarily this portion would not be functional. Deletion of a single nucleotide causes a shift in the "reading frame." Translation of the nucleic acid code proceeds from one end, with three nucleotides at a time being interpreted for the corresponding amino acid. Loss of a nucleotide causes the remaining nucleotides to shift by one position, creating a new sequence of codons subsequent to the deletion. This produces a polypeptide chain which is normal up to a certain point but totally different beyond that.

Two of the 64 possible nucleotide triplets appear to be "punctuation." That is, when they occur, translation into protein stops and the polypeptide is released as "finished." These triplets, UAA and UAG in the ribonucleic acid code, can arise from other codons by single nucleotide substitution in the DNA. If they should arise in the middle of the nucleic acid message, synthesis would be interrupted at that point and incomplete chains would be released.

SO FAR we have said nothing concerning the quantitative control of gene action. Unfortunately, very little is known about this subject in higher organisms. It is clear that some genes work only in some tissues and not in others. Hemoglobin synthesis is an example. In addition, there appear to be mechanisms for informing a gene when it has done enough or when it needs to speed up. The models which have been worked out in bacteria may be valid also for other organisms.

In the special case of human hemoglobin, the rate of synthesis depends in part on the structure. Hemoglobin S is synthesized only about half as fast as normal hemoglobin, and some variants are synthesized at still slower relative rates. It is possible that the main advantage of many molecular forms is due to the rapidity with which they can be synthesized rather than to functional superiority of the completed molecule. The main disadvantage of some hemoglobin variants seems to be a low rate of synthesis resulting in anemia, since the hemoglobins function quite normally in oxygen transport.

Although much of this discussion is in terms of abnormal variation, it is increasingly apparent that many human proteins, perhaps the majority, exist in a variety of inherited forms with normal function. To say that the function is normal is not to imply absence of functional differences. Perhaps the situations which might expose the functional differences do not occur in nature or occur only rarely. On the other hand, normal persons do differ in many ways from each other, and ultimately many of these differences are due to variations in gene complement. Some examples are considered in the section on pharmacogenetics.

THE PREVIOUS section has considered ways in which mutant genes affect proteins. In terms of the physiological consequences, most mutant genes result in diminished or total loss of function of a specific protein. Since most proteins are enzymes which catalyze specific reactions of metabolism, mutations can usually be related to their effect on a certain metabolic reaction.

One of the better known examples of an inherited metabolic defect is the disease phenylketonuria (PKU). This is a typical recessive trait, in that persons with one PKU gene and a normal gene show no ill effects. A person with two PKU genes is nearly always severely mentally defective. The name phenylketonuria comes from the excretion of phenylpyruvic acid (a phenyl ketone) in the urine. The metabolic reactions important in phenylketonuria are shown in Figure 3. Normally, the amino acid phenylalanine is converted to tyrosine, which is oxidized further to recover the energy in the molecule. There are several other reactions of tyrosine of great importance physiologically, although only a small portion of the tyrosine is utilized in these pathways.

The conversion of phenylalanine to tyrosine is an oxidative step catalyzed by the liver enzyme phenylalanine hydroxylase. Phenylketonurics lack this enzyme function and therefore cannot make the conversion. The phenylalanine consumed as protein accumulates, and affected persons may have 50 to 100 times normal blood levels. Because of these high levels, side reactions occur which, under ordinary circumstances, are quantitatively unimportant.

Figure 3. Metabolic pathways important in phenylalanine metabolism. The reactions indicated with broken arrows ordinarily are not important quantitatively. In phenylketonuria (PKU), the normal conversion of phenylalanine to tryosine is blocked, and much of the phenylalanine undergoes reactions indicated by the broken arrows.

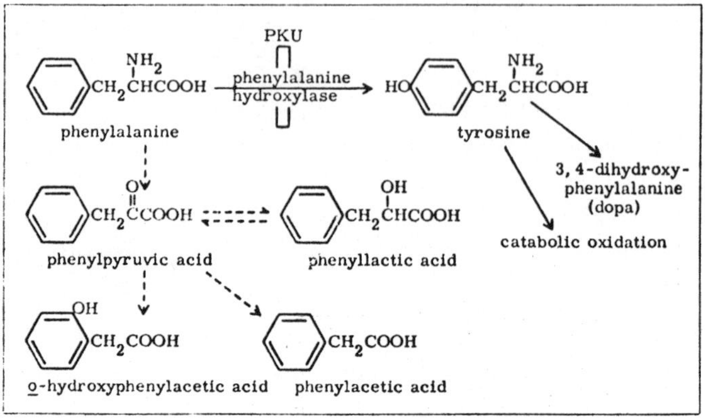

These are responsible for the formation of phenylpyruvic acid, *o*-hydroxyphenylacetic acid, and other substances characteristic of PKU. Some of the products of these side reactions are toxic to the central nervous system and produce irreversible brain damage. It has not been established whether lack of phenylalanine hydroxylase activity results from loss of the protein or from an altered form of the protein which cannot function.

Phenylketonuria also serves as a model for one approach to treatment of inherited defects. Human beings cannot make phenylalanine, relying on dietary protein as a source. It follows that if phenylalanine were removed from the diet, persons with PKU should not be affected. The difficulty is that phenylalanine is required as a building block for body proteins, so that complete elimination would be very detrimental. The amount consumed can be controlled, however, and good results have been obtained in preventing mental defect when proper diets are instituted within the first weeks of life. Delay in starting diet control leads to irreversible changes in the nervous system, hence the institution of large-scale screening programs for detection of PKU in newborns.

Dozens of diseases involving metabolic blocks have been identified in recent years. Each varies in profoundness of the defect and means of therapy, depending on the unique metabolic and physiological aspects of the block. Some

16

are relatively benign, and others may be within the range of what is considered normal variation.

THE TERM pharmacogenetics is applied to inherited variation which is exposed only when the person challenges his metabolism with drugs. The best known example is glucose-6-phosphate dehydrogenase (G6PD) deficiency. This enzyme is widespread in tissues and is important in glucose metabolism. Some persons inherit a form of the enzyme which is less stable than the common form. This is particularly easy to observe in red cells, where synthesis of proteins is completed early in the maturation of the cell and there is no mechanism for replenishment.

Persons with this defect are entirely normal except when challenged with certain drugs or when they eat fava beans. The drugs include the anti-malarial primaquine, which first led to recognition of the condition, as well as a variety of more common substances such as naphthalene mothballs. When exposed to these substances, G6PD-deficient red cells are unable to cope with the chemical stress and break open. The release of hemoglobin into the plasma may have very undesirable consequences, including death.

The genetic locus for G6PD is on the X chromosome. Males, with one X chromosome, are either clearly affected or nonaffected. Females may have one or two G6PD deficiency genes and show intermediate effects.

Another drug handled differently by different persons is isoniazid, used primarily in treatment of tuberculosis. Some persons inactivate the drug rapidly, others slowly, depending on the gene combinations at a specific locus.

An important area, still largely unexplored, concerns inherited variations in resistance to infection. Work with experimental animals indicates clearly the innate differences among inbred strains in ability to cope with infections. The most suggestive studies in human beings concern leprosy, where only a portion of the population appears to be susceptible. Until such differences can be tied to specific genes, the interpretation will be suspect. Nonetheless, this is an area in which significant advances can be anticipated. It also emphasizes the interrelations of heredity and environment, both always contributing to the final makeup of the individual.

17

HUMAN CHROMOSOME ABNORMALITIES AS RELATED TO PHYSICAL AND MENTAL DYSFUNCTION

John H. Heller

THE relationship of human disease syndromes to chromosome aberrations is assuming an increasingly greater role in the detection, diagnosis, treatment and prediction of mental and physical defects in man. By means of karyotype analysis one is enabled to recognize previously unknown syndromes and to differentiate between separate but phenotypically similar entities. Proper diagnosis permits suitable therapeutic measures to be undertaken and enables genetic counselors to assess correct risks in many instances. Recent refinements in sampling embryonic cells by amniocentesis make it feasible to determine, in high risk cases, whether the embryo has a chromosome abnormality or whether it is a male, which has a high risk of sex-linked genetic defect. Termination of pregnancy can be recommended on the basis of this knowledge.

Classes of Chromosome Abnormalities

Chromosome abnormalities have been known in plant and animal species for a very long time. They

Dr. Heller is president of the New England Institute, Ridgefield, Connecticut 06877. This paper was prepared in collaboration with Dr. George H. Mickey, the New England Institute, and was presented on August 19, 1969, as the Sixth Wilhelmine E. Key lecture at the annual meetings of the American Institute of Biological Sciences, University of Vermont, Burlington. The key lecture was established by the American Genetic Association through funds bequeathed to the Association by Dr. Wilhelmine E. Key for the support of lectures in human genetics.

THE JOURNAL OF HEREDITY, 1969, Vol. 60, pp. 239-248.

occur firstly as variations in the number of chromosomes per cell deviating from the normal two sets (maternal and paternal), existing either as complete multiples of sets, a condition called polyploidy (triploidy, tetraploidy, etc.), or as addition or loss of chromosomes within a set, a situation known as aneuploidy (monosomy, trisomy, tetrasomy, etc.). The origin of deviations in chromosome number is known to be through nondisjunction, either during the meiotic divisions in the maturation of the germ cells or during mitotic divisions in the developing individual, or through lagging of chromosomes at anaphase of cell division.

Secondly, chromosome aberrations occur as structural modifications such as duplications, deficiencies, translocations, inversions, isochromosomes, ring chromosomes, etc. These aberrations result from chromosome breakage and reunion in various patterns different from the normal sequence of loci. In most cases, especially the "spontaneous" instances, the cause of chromosome breaks is unknown, but many extraneous agents have been demonstrated experimentally to be efficacious in inducing fragmentation. Foremost among these agents is ionizing radiation but many chemical substances (alkylating agents, nitroso-compounds, antibiotics, DNA precursors, etc.) and viruses have been implicated.

Genetic Effects of Chromosome Aberrations

The striking genetic alterations accompanying chromosome aberrations were brilliantly analyzed by Blakeslee and coworkers on *Datura*, and by the *Drosophila* workers (Morgan, Bridges, Muller, Sturtevant, Painter, Patterson and many others). The task was greatly facilitated in *Drosophila* by the fortunate circumstance in the larval salivary glands where the giant polytene chromosomes exhibit intimate somatic pairing as well as characteristic banding patterns that permit identification of specific gene loci.

Particularly illuminating were Bridges' analyses of sex chromosomes and sex determination in *Drosophila*, utilizing the phenomenon of nondisjuction of the sex chromosomes and culminating in the genic balance theory of sex determination. In this insect the female normally has two X chromosomes plus the autosomes, and the male has one X and one Y. Two X chromosomes and one Y chromosome results in a female, whereas a chromosome constitution of XO produces a sterile male.

In contrast, the Y chromosome in mammals has a strongly masculinizing influence. The presence of a

19

single Y is sufficient to induce differentiation into a male phenotype in the presence of one to five X chromosomes. The XO constitution differentiates into a female phenotype in both mouse and man.

Mammalian Chromosome Studies

The first reported instance of chromosome aberration in mammals was discovered by genetic methods in the waltzing mouse by William H. Gates in 1927[37], and analyzed cytologically by T. S. Painter[68]. Many difficulties in techniques prevented accurate counting and analysis of mammalian chromosomes—large number and relative small size of chromosomes, tendency to clump on fixation, cutting of chromosomes in sectioned material, etc. Even the somatic chromosome number in man was accepted erroneously as 48 until 1956 when Tjio and Levan[88] established the correct count of 46. This count was quickly confirmed by Ford and Hammerton[28], and in 1959 the first positive correlation of a chromosome abnormality and human disease syndrome was made by Lejeune et al.[54] (also Jacobs et al.[44])—the trisomic number 21 chromosome, and Down's syndrome or mongolism. Shortly thereafter Klinefelter's[46] and Turner's[30] syndromes were identified with XXY and XO sex chromosome constitutions respectively, and in rapid succession reports of many other human chromosome abnormalities appeared, such as trisomy 17, trisomy 18, partial trisomy, ring X chromosome, sex chromosome mosaics, cri-du-chat syndrome, et cetera[9, 26].

This sudden explosion of human chromosome studies, in contrast to the long delay of confirmation in human cells of chromosome abnormalities long known in plants and other animals, was made possible by new techniques of preparation. The accumulation of many cells in the metaphase stage of mitosis with colchicine, the use of hypotonic solution to swell the cells and separate chromosomes on the spindle, the discovery that phytohemagglutinin stimulates mammalian peripheral lymphocytes to undergo mitosis, and the method of squashing or spreading on slides of loose cells taken from bone marrow or tissue culture, all contributed to the rapid and accurate analysis of mammalian and human chromosome number and structure.

Karyotype analysis involves the careful comparison of chromosomes in a particular individual to the standard pattern for human cells, including precise measurements of lengths, arm ratios and other morphological features. Special attention is given to comparison of homologous chromosomes

20

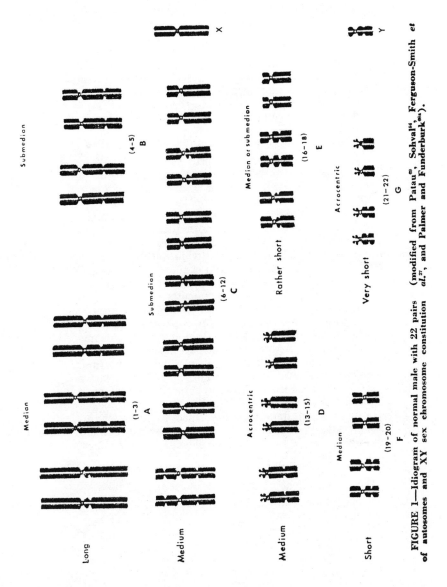

FIGURE 1—Idiogram of normal male with 22 pairs of autosomes and XY sex chromosome constitution (modified from Patau[65], Sohval[85], Ferguson-Smith *et al.*[27], and Palmer and Funderburk[64]).

where differences may indicate abnormalities. An idiogram is a diagrammatic representation of the entire standard chromosome complement, showing their relative lengths, position of centromeres, arm ratios, satellites, secondary constrictions and other features. Figure 1 shows an idiogram of a normal human male with 22 pairs of autosomes and XY sex chromosome constitution. A karyotype is constructed from photographs of chromosomes which are arranged in pairs similar to the idiogram. Figure 2 shows a karyotype of a normal human female.

Incidence of Human Chromosome Anomalies

Chromosome anomalies are relatively frequent events. They have been estimated to occur in 0.48 percent of all newborn infants (one in 208)[81]. At least 25 percent of all spontaneous miscarriages result from gross chromosomal errors[13]. The general incidence of chromosome abnormalities in abortuses is more than fifty times the incidence at birth.

Although it is impossible to obtain an accurate total of victims suffering from effects of chromosome aberrations, one can make rough calculations on the basis of their estimated frequencies in the population of the United States assuming that there is no appreciable difference in life expectancy between these individuals and those with normal chromosome complements. Although this assumption probably is unjustifiable, it suffices for this rough calculation. Among the current population of 202 million we arrive at a figure of 1,136,971 total afflicted with chromosome abnormalities. This total probably represents an underestimate since it does not include all types of chromosome aberrations. Table I indicates totals for a number of specific syndromes.

Syndromes Related to Autosome Abnormalities

Down's syndrome

This defect results from duplication of all or part of autosome 21, either in the trisomic state or as a translocation to another chromosome, usually a 13–15 (D group) or 16–18 (E group) but may be to another G group chromosome. The overall incidence is about 1 in 700 live births[71], but the trisomic type is correlated with age of the mother, having a frequency of about 1 in 2000 in mothers under 30 years of age, and increasing to 1 in 40 in mothers aged 45 or over. The translocation type constitutes about 3.6 percent of cases and is unrelated to the

Table I. Total frequencies in the United States of various types of chromosomal abnormalities, calculated on the basis of 202 million current population and the estimated frequency of each abnormality. (It must be noted that the grand total does not include all types of chromosome aberrations, therefore must be lower than the real value)

Syndrome	Chromosome number	Estimated incidence	Calculated number in U.S.
Down's trisomy 21	47	1 in 700	288,571
Trisomy D	47	1 in 10,000	20,200
Trisomy E	47	1 in 4000	50,500
Trisomy X	47	1 in 10,000 females	101,000
Turner's XO	45	1 in 5000 females	20,200
Klinefelter's XXY	47	1 in 400 males	252,500
Double Y XYY	47	1 in 250 males	404,000
		Total	1,136,971

mother's age, but is transmitted in a predictable manner. Among mental retardates mongoloids represent 16.7 percent.

Clinical features include physical peculiarities ranging from slight anomalies to severe malformations in almost every tissue of the body. Typical appearance of a mongoloid shows slanting eyes, saddle nose, often a large ridged tongue that rolls over a protruding lip, a broad, short skull and thick, short hands, feet and trunk. Frequent complications occur: cataract or crossed eyes, congenital heart trouble, hernias, and a marked susceptibility to respiratory infections. They exhibit characteristic dermatoglyphic patterns on palms and soles. Also they have many biochemical deviations from normal, such as decreased blood-calcium levels and diminished excretion of tryptophane metabolites. Early ageing is common.

All mongoloids are mentally retarded; they usually are 3 to 7 years old mentally. Among the relatively intelligent patients, abstract reasoning is exceptionally retarded.

Female mongoloids are fertile and recorded pregnancies have yielded approximately 50 percent mongoloid offspring. Fortunately male mongoloids are sterile. Examination of their testes reveals varying degrees of spermatogenic arrest correlated with the abnormal chromosome features.

Among mongoloids there is a prevelence of leukemia in childhood; the incidence is some twenty times greater than in the general population. Simultaneous occurrence with other syndromes such as Klinefelter's, also is found, and many cases of mosaicism have been described.

E trisomy syndrome

This is another autosomal anomaly, which involves chromosomes 16, 17 and 18, and is estimated to occur at a frequency of 1 in 4000 live births[20]. Many others die before birth, thus contributing to the large number of miscarriages and stillbirths. These individuals survive only a short time, from one-half day to 1460 days, with an average of 239 days, but females live significantly longer than males.

Trisomy 17 syndrome

Many serious defects usually are present in afflicted invididuals[25]: odd shaped skulls, low-set and malformed ears, triangular mouth with receding chin, webbing of neck, shield-like chest, short stubby fingers, and toes with short nails, webbing of toes, ventricular septal defect and mental retardation, as well as abnormal facies, micrognathia and high arched palate.

Trisomy 18 syndrome

This anomoly[70, 82] is characterized by multiple congenital defects of which the most prominent clinical features are: mental retardation with moderate hypertonicity, low-set malformed ears, small

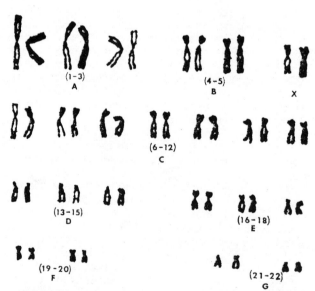

FIGURE 2—Karyotype of a normal human female with 22 pairs of autosomes and two X chromosomes.

mandible, flexion of fingers with the index finger overlying the third, and severe failure to thrive. It generally results in death in early infancy. Its frequency increases with advanced maternal age. Three times as many females as males have been observed; one would expect that more males with this syndrome will be found among stillbirths and fetal deaths.

D syndrome

This trisomy[19, 56, 70, 83] involves chromosomes 13, 14 and 15, and has an estimated frequency of about 1 in 10,000 live births. Many others die in utero. Survival time has been reported from 0 to 1000 days, with an average of 131 days.

Clinical features include: microcephaly, eye anomalies (corneal opacities, colobomata, microphthalmia, anophthalmia), cleft lip, cleft palate, brain anomalies (particularly arrhinencephaly), supernumerary digits, renal anomalies (especially cortical microcysts), and heart anomalies.

Trisomy 22 syndrome

This syndrome produces mentally retarded, schizoid individuals. Reports of its occurrence are too few to permit an estimate of its frequency in the population.

Cri-du-chat syndrome

Lejeune et al.[55] first described this anomaly in 1965, which involves a deficiency of the short arm of a B group chromosome, number 5. Translocations appear to be a common cause of the defect, an estimated 13 percent of cases being associated with translocations; described cases have had B/C, B/G, and B/D translocations[23]. The high proportion emphasizes the importance of unbalanced gamete formation in translocation heterozygotes as a cause of this syndrome. Among parents the frequency of male and female carriers is approximately the same, a situation that contrasts with the much greater frequency of female carriers of a D/G translocation among parents of translocation mongoloids.

Typical clinical features of cri-du-chat individuals are: low birth weight, severe mental retardation, microcephaly, hypertelorism, retrognathism, downward slanting eyes, epicanthal folds, divergent strabismus, growth retardation, narrow ear canals, pes planus and short metacarpals and metatarsals. About 25 to 30 percent of them have congenital heart disorders. A characteristic cat-like cry in

25

infancy is responsible for the name of the syndrome. The cry is due to a small epiglottis and larynx and an atrophic vestibule. However, this major diagnostic sign disappears after infancy, making identification of older cases difficult.

An estimate of the frequency of this syndrome is given as over 1 percent but less than 10 percent of the severely mentally retarded patients. Many have IQ scores below 10, and most are institutionalized.

Philadelphia chromosome

Finally, among autosomal aberrations, a deleted chromosome 21 occurs in blood-forming stem cells in red bone marrow. This deletion, which shows up long after birth, appears to be the primary event causing chronic granulocytic leukemia. This aberration was discussed in 1960 by Nowell and Hungerford[67] (also Baikie et al.[3]).

Syndromes Related to Sex Chromosome Aberrations

The great majority of known chromosomal abnormalities in man involve the sex chromosomes. In one survey (that excluded XYY) it was estimated that abnormalities occurred in 1 out of every 450 births; if the recent estimate of XYY[81] is correct, the frequency actually is much higher. Increased knowledge about sex chromosome aberrations is probably related to the greater concentration of attention on patients with sexual disorders, but is due in part to the ability to detect carriers of an extra X chromosome by the so-called sex chromatin body or Barr body[6]. This structure is a stainable granule at the periphery of a resting nucleus and, according to the Lyon hypothesis[57], is considered to be an inactivated X chromosome. A normal female cell has one Barr body, since it has two X chromosomes, and is said to be sex chromatin positive (or one positive). A normal male cell has no Barr body and is said to be sex chromatin negative.

Klinefelter's syndrome

The first sex chromosome anomaly described in 1959 by Jacobs and Strong[46] and also by Ford et al.[29] was the XXY constitution that is typical of Klinefelter's syndrome. Buccal smears from these patients are sex chromatin positive. They can be tentatively diagnosed by this test along with clinical symptoms. Final confirmation of diagnosis can be achieved by karyotype analysis using either bone marrow aspiration or peripheral blood culture.

Victims of Klinefelter's syndrome are always male

but they are generally underdeveloped, eunachoid in build, with small external genitalia, very small testes and prostate glands, with underdevelopment of hair on the body, pubic hair and facial hair, frequently with enlarged breasts (gynecomastia), and many have a low IQ.

The classical type with two X chromosomes and one Y chromosome was the first case discovered, but subsequently chromosome compositions of XXXY, XXXXY, XXYY[66] and XXXYY[7, 8, 63, 77] have been reported. In addition, numerous mosaics have been described, including double, triple and quadruple numeric mosaics, as well as combinations of numeric and structural mosaics. These conditions are summarized in Table II. They all resemble the XXY Klinefelter's phenotypically and are considered modified Klinefelter's syndromes. The classical XXY type may have low normal mental development or may be retarded, but other types show increasingly greater mental retardation.

The incidence of Klinefelter's syndrome is estimated to be 1 in 400 male live births, which represents from 1 to 3 percent of mentally deficient patients. This condition also has been correlated with age of the mother: the older the mother, the greater the risk of having such a child. These individuals usually are sterile. Spermatogenesis is generally totally absent. Hyalinization of the semeniferous tubules begins shortly before puberty. Congenital malformations are rare. Mental retardation is present in approximately 25 percent of affected individuals, and mental illness may be more common than in the general population.

Turner's syndrome

Female gonadal dysgenesis was described by Turner in 1938 as a syndrome of primary amenorrhea, webbing of the neck, cubitas valgus and short stature, coarctation of aorta, failure of ovarian development and hormonal abnormalities. Patients exhibit sexual infantilism; their breasts are usually underdeveloped, nipples often widely spaced, particularly in those subjects who have a shield or funnel chest deformity. Usually sexual hair is scanty; external genitalia are infantile; labia small or unapparent; clitoris usually normal, although may be enlarged. The uterus is infantile; the tubes long and narrow; the gonads represented by long, narrow, white streaks of connective tissue in normal position of ovary. They are almost always sterile. Hormonal secretions usually are abnormal. Short-

27

Table II. Reported sex chromosomal constitutions in Klinefelter's syndrome (modified from Reitalu[77])

		Sex chromosomal constitution			
Only one karyotype observed per individual		XXY			
		XXYY			
		XXXY			
		XXXYY			
		XXXXY			
Numeric mosaics	Double	XX	XXY		
		XY	XXY		
		XY	XXXY		
		XXY	XXYY		
		XXXY	XXXXY		
		XXXX	XXXXY		
	Triple	XY	XXY	XXYY	
		XX	XXY	XXXY	
		XY	XXY	XXXY	
		XO	XY	XXY	
		XX	XY	XXY	
		XXXY	XXXXY	XXXXYY	
		XXXY	XXXXY	XXXXXY	
	Quadruple	XXY	XY	XX	XO
Numeric and structural mosaics	Double	XXY	XXxY		
	Triple	XY	XXY	XXxY	
		XxY	Xx	XY	

ness of stature is characteristic and many other skeletal abnormalities occur. Peculiar facies include small mandible, anti-mongolian slant of eyes, depressed corners of mouth, low-set ears, auricles sometimes deformed. Cardiovascular defects are frequent, the most common being coarctation of the aorta. Slight intellectual impairment is found in some patients, particularly those with webbing of the neck.

In 1954 it was discovered that many patients with ovarian agenesis were sex chromatin negative, and in 1959 Ford and colleagues[30] gave the first chromosome analysis showing that Turner's syndrome has the sex chromosome abnormality of only one X chromosome (XO) rather than two X's. It was quickly confirmed by Jacobs and Keay[45] and by Fraccaro et al[33].

Mosaicism is known to exist—both 45 chromosome cells and 46 chromosome cells occur side by side in tissues of the individual—and can result from nondisjunction in early embryonic development. Isochromosomes sometimes are involved, e.g., creating a situation with 3 long arms of the X chromosome but only 1 short arm.

The incidence of XO Turner's syndrome is estimated as 1 in approximately 5000 women; many die in utero.

Large scale screening of newborn babies by buccal smears can permit detection of chromatin negative females, chromatin positive males, and double, triple, quadruple and quintuple positive cases of either sex[58]. Table III shows the relationship between sex chromosome complements and sex chromatin pattern.

Triplo-X syndrome

Females containing three[47], four and five X chromosomes are known[4, 47, 63]. The triplo-X syndrome is thought to have an incidence of about 1 in 800 live female births. This syndrome was first described by Jacobs et al.[47] in 1959. Although it has no distinctive clinical picture, menstrual irregularities may be present, secondary amenorrhea or premature menopause. Most cases have no sexual abnormalities and many are known to have children. The most characteristic feature of 3X females is mental retardation. Qudruple-[14] and quintuple-X[50] syndromes are much rarer. These individuals are mentally retarded, usually the more X chromosomes present, the more severe the retardation. Frequently these individuals are fertile.

An extra X chromosome confers twice the usual risk of being admitted to a hospital with some form of mental illness. The loss of an X, on the other hand, has no association with mental illness; thus the chance of mental hospital admission is not raised for an XO female. An extra X chromosome also predisposes to mental subnormality. The prevalence of psychosis among patients in hospitals for the subnormal is unusually high in males with two or more X chromosomes.

Numerous other sex chromosome anomalies occur[38], many involving mosaics and structural chromosome aberrations. For example, occasionally an XY embryo will differentiate into a female, a situation referred to as testicular feminization male pseudohermaphrodite (Morris syndrome)[60]. These individuals have only streak gonads and vestigial internal genital organs. They usually have undeveloped breasts and do not menstruate. They are invariably sterile[76].

Still other sexual abnormalities are intersexes and true hermaphrodites, many of which have an XX sex chromosome constitution or are mosaics for sex chromosomes such as XO/XY or XX/XY or more complicated mixtures[31]. Sex chromosome mosaicism is very common. Almost every sex chromosome combination found alone has been found in association with one or more cell lines with a different sex chromosome constitution. These mosaics exhibit quite a variable expression; for example in an XO/XY mosaic the external genitalia can appear female, male or intersexual[85].

The YY syndrome

The male with an extra Y chromosome (XYY) has attracted much attention in the public press as well as in scientific circles because of his reputed antisocial, aggressive and criminal tendencies[1, 2, 64]. Although this abnormality belongs in the above category of syndromes related to sex chromosome aberrations, it has been singled out for special discussion because of its social and legal implications.

Evidence supporting the existence of a double Y syndrome has accumulated within the last six years. Studies in Sweden[32] showed an unusually large number of XXYY and XYY men among hard-to-manage patients in mental hospitals. These observations received impressive confirmation in studies of maximum security prisons and hospitals for the criminally insane in Scotland where an astonishingly high frequency (2.9 percent) of XYY males were found[48]. This was over fifty times higher than the then current estimate of 1 in 2000 in the general population. Subsequently many additional studies on the YY syndrome have appeared and a composite picture of the XYY male emerged[5, 19, 15, 21a, 34, 35, 41, 72-75, 78, 80, 92].

The principal features of the extra Y syndrome appear to be exceptional height and a serious personality disorder leading to behavioral disturbances. It seems likely it is the behavior disorder rather than their intellectual incompetence that prevents them from functioning adequately in society[18].

Clinically the XYY males are invariably tall (usually six feet or over) and frequently of below-average intelligence. They are likely to have unusual sexual tastes, often including homosexuality. A history of antisocial behavior, violence and conflict with the police and educational authorities from early years is characteristic[86] of the syndrome.

Although these males usually do not exhibit obvious physical abnormalities[12, 24, 40, 42, 52, 91], several cases of hypogonadism[11], some with undescended testes, have been reported. Others have epilepsy,

30

malocclusion and arrested development[87], but these symptoms may be fortuitously associated. One case was associated with trisomy 21[61], another with pseudohermaphrodism[35]. The common feature of an acne-scarred face may be related to altered hormone production. The criminally aggressive group were found to have evidence of an increased androgenic steroid production as reflected by high plasma and urinary testosterone levels[12, 43]. If the high level of plasma testosterone is characteristic of XYY individuals, it suggests a mechanism through which this condition may produce behavioral changes, possibly arising at puberty.

Antisocial and aggressive behavior in XYY individuals may appear early in life, however, as evidenced by a case reported by Cowie and Kahn[22]. A prepubertal boy with normal intelligence, at the age of 4½ years, was unmanageable, destructive, mischievous and defiant, overadventurous and without fear. His moods alternated; there were sudden periods of overactivity at irregular intervals when he would pursue his particular antisocial activity with grim intent. Between episodes he appeared happy and constructive. The boy was over the 97th percentile in height for his age, a fact that supports the view that increased height in the XYY syndrome is apparent before puberty.

It has been suggested that the ordinary degree of aggressiveness of a normal XY male is derived from his Y chromosome, and that by adding another

Table III. The relationship between sex, sex chromosome complement and sex chromatin pattern (modified from Miller[63])

Sex chromatin pattern	Sexual phenotype	
	Female	Male
−	XO	XY
−	XY	XYY
	(testicular feminization)	
+	XX	XXY
		XXYY
++	XXX	XXXY
		XXXYY
+++	XXXX	XXXXY
++++	XXYYX	

Y a double dose of those potencies may facilitate the development of aggressive behavior[65] under certain conditions. A triple dose (XYYY) would be present in the case reported by Townes et al.[90].

The first reported case of an XYY constitution[39, 79] was studied because the patient had several abnormal children, although he appeared to be normal himself. Until recently, reports of the XYY constitution have been uncommon, probably because no simple method exists for screening the double-Y condition that is comparable to the buccal smear—sex chromatin body technique for detecting an extra X chromosome. Another possible explanation for the rarity of reports on the XYY karyotype is the absence of a specific phenotype in connection with it. Most syndromes with a chromosome abnormality are ascertained because of some symptom or clinical sign that indicates a need for chromosome analysis. Consequently there have been few studies that place the incidence of this chromosome abnormality in its proper perspective to the population as a whole.

Very recently a study of the karyotypes of 2159 infants born in one year was made by Sergovich et al.[81]. These investigators detected 0.48 percent of gross chromosome abnormalities. In this sample the XYY condition appeared in the order of 1 in 250 males, which would make it the most common form of aneuploidy known for man. The previous estimate was about 1 in 2000 males. If this figure of 1/250 is valid for the population as a whole, it means that the great majority of cases go undetected and consequently must be phenotypically normal and behave near enough to the norm to go unrecognized.

Several cases of asymptomatic males have been published, including the first one described (Sandberg et al.[79] and Hauschka et al.[39]), which proved to be fertile. It appears that the sons of XYY men do not inherit their father's extra Y chromosome[59a].

Another fertile XYY male, reported by Leff and Scott[53], had inferiority feelings, was slightly hypochondriacal and obsessional, and not very aggressive. He gave a general impression of emotional immaturity. He was 6 feet, 6 inches tall, healthy, with normal genitalia and electroencephalogram. His IQ was 118. Wiener and Sutherland[93] discovered by chance an XYY male who was normal; he was 5 feet, 9½ inches tall, with normal genitalia and body hair, normal brain waves, and with an IQ of 97. He exhibited a cheerful disposition and mild temperament, had no apparent behavioral disturbance and never required psychiatric advice. This case supports the idea that an XYY male can lead a normal life.

Social and Legal Implications of the YY Syndrome

The concept that when a human male receives an

extra Y chromosome it may have an important and potentially antisocial effect upon his behavior is supported by impressive evidence[15, 21]. Lejuene states that "There are no born criminals but persons with the XYY defect have considerably higher chances." Price and Whatmore[74] describe these males as psychopaths, "unstable and immature, without feeling or remorse, unable to construct adequate personal relationships, showing a tendency to abscond from institutions and committing apparently motiveless crimes, mostly against property." Casey and coworkers[16] examined the chromosome complements in males 6 feet and over in height and found: 12 XYY among 50 mentally subnormal and 4 XYY among mentally ill patients detained because of antisocial behavior; also 2 XYY among 24 criminals of normal intelligence. They concluded that their results indicate that an extra Y chromosome plays a part in antisocial behavior even in the absence of mental subnormality. The idea that criminals are degenerates because of bad heredity has had wide appeal. There is no doubt that genes do influence to some extent the development of behavior. The influence may be strongly manifested in some cases but not in others. Some individuals appear to be driven to aggressive behavior.

Several spectacular crime cases served to publicize this genetic syndrome, and it has been played up in newspapers, news magazines, radio and television. In 1965 Daniel Hugon, a stablehand, was charged with the murder of a prostitute in a cheap Paris hotel. Following his attempted suicide he was found to have an XYY sex chromosome constitution. Hugon surrendered to the police and his lawyers contended that he was unfit to stand trial because of his genetic abnormality. The prosecution asked for five to ten years; the jury decided to give him seven.

Richard Speck, the convicted murderer of eight nurses in Chicago in 1966, was found to have an XYY sex chromosome constitution. He has all the characteristics of this syndrome found in the Scottish survey: he is 6 feet 2 inches tall, mentally dull, being semiliterate with an IQ of 85, the equivalent of a 13-year-old boy. Speck's face is deeply pitted with acne scars. He has a history of violent acts against women. His aggressive behavior is attested by his record of over 40 arrests. Speck was sentenced to death but the execution has been held up pending an appeal of the conviction.

In Melbourne, Australia, Lawrence Edward Hannell, a 21-year-old laborer on trial for the stabbing of a 77-year-old widow, faced a maximum sentence of death. He was found to have an XYY constitution, mental retardation, an aberrant brain wave pattern, and a neurological disorder. Hannel pleaded not guilty by reason of insanity, and a criminal court jury found him not guilty on the ground that he was insane at the time of the crime.

A second Melbourne criminal with an XYY constitution, Robert Peter Tait, bludgeoned to death an 81-year-old woman in a vicarage where he had gone seeking a handout. He was convicted of murder and sentenced to hang, but his sentence was commuted to life imprisonment.

Another case is that of Raymond Tanner, a convicted sex offender, who pleaded guilty to the beating and rape of a woman in California. He is 6 feet 3 inches tall, mentally disordered, and has an XYY complement. A superior court judge is attempting to decide whether Tanner's plea of guilty to assault with intent to commit rape will stand, or whether he will be allowed to plead innocent by reason of insanity.

Criminal lawyers in the United States have already begun to request genetic studies of their clients. In October of 1968 a lawyer for Sean Farley, a 26-year-old XYY man in New York who was charged with a rape-slaying, maneuvered to raise the issue of his client's genetic defect in court.

Many questions are raised by the double Y syndrome—basic social, legal and ethical questions—which will become more and more insistent as the implications of chromosome abnormalities take root in the public mind. Is an extra Y chromosome causally related to antisocial behavior? Is there a genetic basis for criminal behavior? If a man has an inborn tendency toward criminal behavior, can we fairly hold him legally accountable for his acts? If a criminal's chromosomes are at fault, how can we rehabilitate him?

The evidence to date is inadequate to prove conclusively the validity of the syndrome and convict all of the world's estimated five million XYY males of innate aggressive or criminal tendencies. But if the concept is proved, what then? The first step would seem to be to identify the XYY infants in the general population. This suggests the need for a nationwide program of automatic chromosome analysis of all newborns.

How should society deal with XYY individuals? If they are genetically abnormal, they should not be treated as normal. If the XYY condition dooms a man to a life of crime, he should be restrained but not punished. Mongolism also is a chromosome

34

abnormality, and afflicted individuals are not held responsible for their behavior. Some valuable suggestions on the legal aspects of the double Y syndrome have been published recently by Kennedy McWhirter[59]. Elsewhere, Kessler and Moos[51] claim that definitive concepts relating to the YY syndrome have been accepted prematurely.

If all infants could be karyotyped at birth or soon after, society could be forearmed with information on chromosome abnormalities and perhaps it could institute the proper preventive and other measures at an early age. Although society can not control the chromosomes (at least at the present time) it can do a great deal to change certain environmental conditions that may encourage XYY individuals to commit criminal acts.

The theory that a genetic abnormality may predispose a man to antisocial behavior, including crimes of violence, is deceptively and attractively simple, but will be difficult to prove. Extensive chromosome screening with prospective follow-up of XYY males will be essential to determine the precise behavioral risk of this group. It is by no means universally accepted yet. Many geneticists urge that we should be cautious in accepting the interpretation that the double Y condition is specifically associated with criminal behavior, and particularly so with reference to the medicolegal validity of these concepts.

Literature Cited

1. ANONYMOUS. The YY syndrome. *Lancet* 1:583–584. 1966.
2. ————. Criminal behavior—XYY criterion doubtful. *Science News* 96:2. 1969.
3. BAIKIE, A. G., W. M. COURT BROWN, K. E. BUCKTON, D. G. HARNDEN, P. A. JACOBS, and J. M. TOUGH. A possible specific chromosome abnormality in human chronic myeloid leukemia. *Nature* 188:1165–1166. 1960.
4. ————, O. MARGARET GARSON, SANDRA M. WESTE, and JEAN FERGUSON. Numerical abnormalities of the X chromosome. *Lancet* 1:398–400. 1966.
5. BALODIMOS, MARIOS C., HERMANN LISCO, IRENE IRWIN, WILMA MERRILL, and JOSEPH F. DINGMAN. XYY karyotype in a case of familial hypogonadism. *J. Clin. Endocr.* 26:443–452. 1966.
6. BARR, M. L. Sex chromatin and phenotype in man. *Science* 130:679. 1959.
7. ———— and D. H. CARR. Sex chromatin, sex chromosomes and sex anomalies. *Canad. Med. Assn. J.* 83:979–986. 1960.
8. ————, D. H. CARR, H. C. SOLTAN, RUTH G. WIENS, and E. R. PLUNKETT. The XXYY variant of Klinefelter's syndrome. *Canad. Med. Assn. J.* 90:575–580. 1964.
9. BELSKY, JOSEPH L. and GEORGE H. MICKEY. Human cytogenetic studies. *Danbury Hospital Bull.* 1:19–20. 1965.

10. Boczkowski, K. and M. D. Casey. Pattern of DNA replication of the sex chromosomes in three males, two with XYY and one with XXYY karyotype. *Nature* 213:928-930. 1967.

11. Buckton, Karin.E., Jane A. Bond, and J. A. McBride. An XYY sex chromosome complement in a male with hypogonadism. *Human Chromosome Newsletter* No. 8, p. 11. Dec. 1962.

12. Carakushansky, Gerson, Richard L. Neu, and Lytt I. Gardner. XYY with abnormal genitalia. *Lancet* 2:1144. 1968.

13. Carr, D. H. Chromosome studies in abortuses and stillborn infants. *Lancet* 2:603-606. 1963.

14. ———, M. L. Barr, and E. R. Plunkett. An XXXX sex chromosome complex in two mentally defective females. *Canad. Med. Assn. J.* 84:131-137. 1961.

15. Casey, M. D., C. E. Blank, D. R. K. Street, L. J. Segall, J. H. McDougall, P. J. McGrath, and J. L. Skinner. YY chromosomes and antisocial behavior. *Lancet* 2:859-860. 1966.

16. ———, L. J. Segall, D. R. K. Street, and C. E. Blank. Sex chromosome abnormalities in two state hospitals for patients requiring special security. *Nature* 209:641-642. 1966.

17. ———, D. R. K. Street, L. J. Segall, and C. E. Blank. Patients with sex chromatin abnormality in two state hospitals. *Ann. Human Genet.* 32:53-63. 1968.

18. Close, H. G., A. S. R. Goonetilleke, Patricia A. Jacobs, and W. H. Price. The incidence of sex chromosomal abnormalities in mentally subnormal males. *Cytogenetics* 7:277-285. 1968.

19. Conan, P. E. and Bayzar Erkman. Frequency and occurrence of chromosomal syndromes. I. D-trisomy. *Am. J. Human Genet.* 18:374-386. 1966.

20. ——— and ———. Frequency and occurrence of chromosomal syndromes. II. E-trisomy. *Am. J. Human Genet.* 18:387-398. 1966.

21. Court Brown, W. M. Sex chromosomes and the law. *Lancet* 2:508-509. 1962.

21a. ———. Males With an XYY sex chromosome complement. *J. Med. Genet.* 5:341-359. 1968.

22. Cowie, John and Jacob Kahn. XYY constitution in prepubertal child. *Brit. Med. J.* 1:748-749. 1968.

23. deCapoa, A., D. Warburton, W. R. Breg, D. A. Miller, and O. J. Miller. Translocation heterozygosis: a cause of five cases of *cri du chat* syndrome and two cases with a duplication of chromosome number five in three families. *Am. J. Human Genet.* 19:586-603. 1967.

24. Dent, T., J. H. Edwards, and J. D. A. Delhanty. A partial mongol. *Lancet* 2:484-487. 1963.

25. Edwards, J. H., D. G. Harnden, A. H. Cameron, V. Mary Crosse, and O. H. Wolff. A new trisomic syndrome. *Lancet* 1:787-790. 1960.

26. Eggen, Robert R. Chromosome Diagnostics in Clinical Medicine. Charles C. Thomas, Springfield, Ill. 1965.

27. Ferguson-Smith, M. A., Marie E. Ferguson-Smith, Patricia M. Ellis, and Marion Dickson. The sites and relative frequencies of secondary constrictions in human somatic chromosomes. *Cytogenetics* 1:325-343. 1962.

28. Ford, C. E. and J. L. Hamerton. The chromosomes of man. *Nature* 178:1020-1023. 1956.

29. ———, K. W. Jones, O. J. Miller, Ursula Mittwoch, L. S. Penrose, M. Ridler, and A. Shapiro. The chromosomes in a patient showing both mongolism and the Klinefelter syndrome. *Lancet* 1:709-710. 1959.

30. ———, K. W. Jones, P. E. Polani, J. C. deAlmedia, and J. H. Briggs. A sex-chromosome anomaly in a case of gonadal dysgenesis (Turner's syndrome). *Lancet* 1:711–713. 1959.

31. ———, P. E. Polani, J. H. Briggs, and P. M. F. Bishop. A presumptive human XXY/XX mosaic. *Nature* 183:1030–1032. 1959.

32. Forssman, H. and G. Hambert. Incidence of Klinefelter's syndrome among mental patients. *Lancet* 1:1327. 1963.

33. Fraccaro, M., K. Kaijser, and J. Lindsten. Chromosome complement in gonadal dysgenesis (Turner's syndrome). *Lancet* 1:886. 1959.

34. ———, M. Glen Bott, P. Davies, and W. Schutt. Mental deficiency and undescended testia in two males with XYY sex chromosomes. *Folia Hered. Pathol.* (Milan) 11:211–220. 1962.

35. Franks, Robert C., Kenneth W. Bunting, and Eric Engel. Male pseudohermaphrodism with XYY sex chromosomes. *J. Clin. Endocr.* 27:1623–1627. 1967.

36. Fraser, J. H., J. Campbell, R. C. MacGillivray, E. Boyd, and B. Lennox. The XXX syndrome—frequency among mental defectives and fertility. *Lancet* 2:626–627. 1960.

37. Gates, William H. A case of non-disjunction in the mouse. *Genetics* 12:295–306. 1927.

38. Hamerton, J. L. Sex chromatin and human chromosomes. *Intern. Rev. Cytol.* 12:1–68. 1961.

39. Hauschka, Theodore S., John E. Hasson, Milton N. Goldstein, George F. Koepf, and Avery A. Sandberg. An XYY man with progeny indicating familial tendency to non-disjunction. *Am. J. Human Genet.* 14:22–30. 1962.

40. Hayward, M. D. and B. D. Bower. Chromosomal trisomy associated with the Sturge-Weber syndrome. *Lancet* 2:844–846. 1960.

41. Hunter, H. Chromatin-positive and XYY boys in approved schools. *Lancet* 1:816. 1968.

42. Hustinx, T. W. J. and A. H. F. van Olphen. An XYY chromosome pattern in a boy with Marfan's syndrome. *Genetica* 34:262. 1963.

43. Ismail, A. A. A., R. A. Harkness, K. E. Kirkham, J. A. Loraine, P. B. Whatmore, and R. P. Brittain. Effect of abnormal sex-chromosome complements on urinary testosterone levels. *Lancet* 1:220–222. 1968.

44. Jacobs, P. A., A. G. Baikie, W. M. Court Brown, and J. A. Strong. The somatic chromosomes in mongolism. *Lancet* 1:710. 1959.

45. ——— and A. J. Keay. Chromosomes in a child with Bonnevie-Ullrich syndrome. *Lancet* 2:732. 1959.

46. ——— and J. A. Strong. A case of human intersexuality having a possible XXY sex-determining mechanism. *Nature* 182:302–303. 1959.

47. ———, A. G. Baikie, W. M. Court Brown, T. N. MacGregor, N. MacLean, and D. G. Harnden. Evidence for the existence of the human "super female". *Lancet* 2:423–425. 1959.

48. ———, Muriel Brunton, Marie M. Melville, R. P. Brittain, and W. F. McClemont. Aggressive behaviour, mental subnormality and the XYY male. *Nature* 208:1351–1352. 1965.

49. ———, W. H. Price, W. M. Court Brown, R. P. Brittain, and P. B. Whatmore. Chromosome studies on men in a maximum security hospital. *Ann. Human Genet.* 31:330–347. 1968.

50. Kesaree, Nirmala and Paul V. Woolley. A phenotypic female with 49 chromosomes, presumably XXXXX. *J. Pediat.* 63:1099–1103. 1963.

51. KESSLER, SEYMOUR and RUDOLPH H. Moos. XYY chromosome: premature conclusions. *Science* 165:442. 1969.

52. KOSENOW, W. and R. A. PFEIFFER. YY syndrome with multiple malformations. *Lancet* 1:1375-1376. 1966.

53. LEFF, J. P. and P. D. SCOTT. XYY and intelligence. *Lancet* 1:645. 1968.

54. LEJEUNE, J., M. GAUTIER, and R. TURPIN. Etude des chromosomes somatiques de neuf enfants mongoliens. *Compt. Rend. Acad. Sci.* 248:1721-1722. 1959.

55. ————, J. LAFOURCADE, R. BERGER and M. O. RETHORE. Maladie du cri du chat et sa reciproque. *Ann. Genet.* 8:11-15. 1965.

56. LUBS, H. A., JR., E. V. KOENIG, and L. H. BRANDT. Trisomy 13-15: A clinical syndrome. *Lancet* 2:1001-1002. 1961.

57. LYON, M. F. Gene action in the X-chromosome of the Mouse (*Mus musculus* L.). *Nature* 190:372-373. 1961.

58. MACLEAN, N., D. G. HARNDEN, W. M. COURT BROWN, JANE BOND, and D. J. MANTLE. Sex-chromosome abnormalities in newborn babies. *Lancet* 1:286-290. 1964.

59. MCWHIRTER, KENNEDY. XYY chromosome and criminal acts. *Science* 164:1117. 1969.

59a. MELNYK, JOHN, FRANK VANASEK, HAVELOCK THOMPSON, and ALFRED J. RUCCI. Failure of transmission of supernumerary Y chromosomes in man. Abst. Am. Soc. Human Genet. Annual Meeting, Oct. 1-4, 1969.

60. MICKEY, GEORGE H. Chromosome studies in testicular feminization syndrome in human male pseudohermaphrodites. *Mammalian Chromosome Newsletter* No. 9, p. 60. 1963.

61. MIGEON, BARBARA R. G trisomy in an XYY male. *Human Chromosome Newsletter* No. 17. Dec. 1965.

62. MILCU, M., I. NIGOESCU, C. MAXIMILIAN, M. GAROIU, M. AUGUSTIN, and ILEANA ILIESCU. Baiat cu hipospadias si cariotip XYY. *Studio si Cercetari de Endocrinologie* (Bucharest) 15:347-349. 1964.

63. MILLER, ORLANDO J. The sex chromosome anomalies. *Am. J. Obstet. Gynec.* 90:1078-1139. 1964.

64. MINCKLER, LEON S. Chromosomes of criminals. *Science* 163:1145. 1969.

65. MONTAGU, ASHLEY. Chromosomes and crime. *Psychology Today* 2:43-49. 1968.

66. MULDAL, S. and C. H. OCKEY. The "double male": a new chromosome constitution in Klinefelter's syndrome. *Lancet* 2:492-493. 1960.

67. NOWELL, P. C. and D. A. HUNGERFORD. A minute chromosome in human granulocytic leukemia. *Science* 132:1497. 1960.

68. PAINTER, T. S. The chromosome constitution of Gates "non-disjunction" (v-o) mice. *Genetics* 12:379-392. 1927.

68a. PALMER, CATHERINE G. and SANDRA FUNDERBURK. Secondary constrictions in human chromosomes. *Cytogenetics* 4:261-276. 1965.

69. PATAU, K. The identification of individual chromosomes, especially in man. *Am. J. Human Genet.* 12:250-276. 1960.

70. ————, D. W. SMITH, E. THERMAN, S. L. INHORN, and H. P. WAGNER. Multiple congenital anomalies caused by an extra chromosome. *Lancet* 1:790-793. 1960.

71. PENROSE, L. S. The Biology of Mental Defect. Grune and Stratton, New York. 1949.

72. PERGAMENT, EUGENE, HIDEO SATO, STANLEY BERLOW, and RICHARD MINTZER. YY syndrome in an American negro. *Lancet* 2:281. 1968.

73. PFEIFFER, R. A. Der Phanotyp der Chromosomenaberration XYY. *Wochenschrift* 91:1355-1256. 1966.

74. PRICE, W. H. and P. B. WHATMORE. Behaviour disorders and the pattern of crime among XYY males identified at a maximum security hospital. *Brit. Med. J.* 1:533. 1967.

75. ———, J. A. STRONG, P. B. WHATMORE, and W. F. McCLEMENT. Criminal patients with XYY sex-chromosome complement. *Lancet* 1:565–566. 1966.

76. PUCK, T. T., A. ROBINSON, and J. H. TJIO. A familial primary amenorrhea due to testicular feminization. A human gene affecting sex differentiation. *Proc. Exper. Biol. Med.* 103:192–196. 1960.

77. REITALU, JUHAN. Chromosome studies in connection with sex chromosomal deviations in man. *Hereditas* 59:1–48. 1968.

78. RICCI, N. and P. MALACARNE. An XYY human male. *Lancet* 1:721. 1964.

BIOCHEMICAL AND CHROMOSOMAL DEFECTS YIELD TO PRENATAL DIAGNOSIS

by Paul T. Libassi

In the past, the best information that a clinician could provide to prospective parents was a prediction of whether or not their child was likely to be normal. In essence, odds were quoted. Today some 24 inherited biochemical defects as well as all of the major chromosomal abnormalities can be diagnosed prenatally with near certainty.

Amniocentesis—the extraction of fluid which surrounds the fetus—is the keystone of prenatal diagnosis, providing the raw materials from which diagnosis proceeds. Fetuses shed some cells into the amniotic fluid in which they are bathed. These cells can be retrieved as early as the eleventh week of fetal life and grown in culture. Subsequently, biochemical assays for enzyme deficiencies can be performed and chromosomes can be visualized to permit diagnosis of the various chromosomal aberrations.

Timing in dispute. The optimal time during pregnancy when diagnostic amniocentesis is most safely performed remains a matter of dispute. A compromise must be reached between the conflicting facts that the later an amniocentesis is performed, the safer it is, and the earlier that abortion occurs, the safer it is.

Dr. Fritz Fuchs, Chairman of the Department of Obstetrics and Gynecology at Cornell University Medical College maintains that during the 15th week of pregnancy sufficient amounts of amniotic fluid are available to permit the removal of 10-20 ml of fluid without causing adverse effects in either fetus or mother. Amniocentesis performed during the 15th week of gestation assumes that abortion should be performed no later than the 20th week and allows four to five weeks for cells to grow in culture and for diagnosis to be made. If biochemical or chromosomal determinations can be completed in less than four weeks time, amniocentesis should then be delayed for several weeks until the 16th or even 17th week of pregnancy.

Cells harvested from amniotic fluid are cultured for analysis—either biochemical or chromosomal—according to the following general scheme: Fetal cells suspended in the amniotic fluid are separated through centrifugation and, after all liquid has been drained, are placed in a culture medium. In the case of biochemical analysis, cells are grown in this culture medium until sufficient amounts are available for enzyme deficiency tests. Cells are harvested for chromosomal analysis when sufficient numbers reach the metaphase stage of division—the point at which each chromosome has divided into two mirror-

LABORATORY MANAGEMENT, 1972, Vol. 10, No. 9, pp. 20-24.

image halves but are still connected at the centromere.

The cells are run through a centrifuge again to separate them from the culture medium and are treated with colchicine to halt mitosis. A dilute salt solution is then added to the cell preparations to induce swelling and to cause the chromosomes to migrate apart. At this point the cells are fixed and stained and a chromosome spread is photographed through the microscope.

Prenatal diagnosis of biochemical defects represents, according to Dr. John W. Littlefield, Chief of the Genetics Unit at Massachusetts General Hospital, an unusual situation in medicine. The diagnostic information which provides the basis for any therapeutic measures to be taken is based upon a single laboratory test for an enzyme deficiency. Because of this, laboratories must take inordinate precautions to develop successful cultures and enlist the aid of confirmatory procedures whenever possible. The preparation of successful cultures depends in great part on the amount of amniotic fluid from which cells are havested. Opinions on just how much fluid should be drawn vary.

Dr. M. Neil Macintyre's laboratory at Case Western Reserve University obtains as much as 40 cc of fluid, a quantity large enough to maintain cultures in as many as six or seven flasks. According to Dr. Macintyre, several important advantages result from multiple culture preparations. The possibility of an erroneous diagnosis due to maternal cell contamination is greatly minimized, and the likelihood of an earlier diagnosis is enhanced by the growth of a greater volume of cells.

Culture methods vary from laboratory to laboratory. A choice of culture techniques should ultimately depend upon those methods which supply the most accurate data in the shortest pos-

sible time, according to Dr. Macintyre. The cytogenetic laboratory is responsible for production of cultures to be utilized for biochemical analysis and for chromosomal preparations of sufficient volume and quality to facilitate a speedy and accurate diagnostic evaluation. Dr. Littlefield's laboratory has found that a medium supplemented with 15 percent calf serum offers the best results. Cultures at his Massachusetts laboratory are re-fed twice a week, a procedure which appears to enhance the growth of even single cells.

Problems. Several types of cells from epithelial-like to fibroblast-like are present in amniotic fluid. Problems arise because clinicians cannot be certain that in the case of female fetuses the cells that have been cultured are fetal and not maternal. Researchers have suggested that several developments would alleviate this problem and confirm any discrepancies.

Specific and clear-cut differentiation of the isozyme content of fetal and adult cells would be of important practical value. Studies have documented differences in the activity of enzymes between cultured amniotic cells and adult skin fibroblasts. In addition, Dr. Littlefield suspects that different isozyme patterns, in addition to quantitative changes in enzyme activity, are displayed by amniotic cells. If an isozyme pattern peculiar to fetal cells could be defined, it would provide an important confirmatory adjunct. Dr. Littlefield adds that such an isozyme pattern should be detectable in a spread of relatively few cells so that confirmation could be made in time to permit drawing a second specimen of amniotic fluid if necessary.

An alternative to specific isozyme patterns to confirm fetal cells is visualization of the fetus before or during amniocentesis, thereby guaranteeing that fetal cells have in fact been ob-

41

tained. Several scientists are molding fiber glass optics into an instrument called a "fetal amnioscope" just for such purposes.

Dr. Henry Nadler, Chairman of the Department of Pediatrics at Northwestern University and a pioneer in amniocentesis, reports that bioengineers at Northwestern have developed an instrument that can be inserted into a needle only slightly larger than the size presently used for amniocentesis. This instrument has facilitated visualization of an increased number of external morphologic anomalies and in the future may allow clinicians to actually remove a snip of fetal tissue assuring that cells being cultured are in fact fetal cells.

Refinements. In standard chromosomal analysis, the photomicrograph is enlarged and the chromosome images are cut out and arranged on a white card into 22 pairs of homologous chromosomes plus the two sex chromosomes. Pairs are arranged in sequential order according to morphological peculiarities based upon size, shape and ratio of arm length.

Several refinements of this manual procedure have since rendered the construction and examination of the karyotype both less time-consuming and more objective. Dr. Robert S. Ledley of the National Biomedical Research Foundation has developed a computer program which can analyze a photomicrograph with complete accuracy in less than 40 seconds.

The computer is programmed to count the total number of chromosomes in each cell and to measure and notate their morphological features. The computer then matches chromosome pairs according to these features and classifies each pair according to a standard karyotype sequence. In this manner, the computer is capable of evaluating small

but clinically significant chromosomal abnormalities which are often undiscernible by the human eye. Dr. Peter W. Neurath of New England Medical Center, Boston, also has developed a computerized chromosome analysis system (*Lab. Mgt.*, Sept. 1970).

Dr. Ledley is presently developing a special-purpose computer that will automate the entire procedure from selection to analysis. Such a system would eliminate any degree of subjectivity involved in the selection of cells to be photographed for analysis. A scanner linked directly to the microscope would automatically select and photograph the appropriate cells. The only problem with the system at this point seems to be the costs involved. Dr. Harold Nitowsky of the Albert Einstein Medical Center explains, "In terms of cost analysis, the manual procedures are certainly more feasible at this point."

The discovery of banding patterns unique for each individual pair of human chromosomes has resulted in a reliable and accurate method for identifying and aligning chromosomes on a karyotype. Dr. Margery Shaw at the University of Texas has found that chromosomes stained with a Giemsa stain display discrete and highly specific banding patterns in characteristic sections of the chromosome arm. Not only does this technique permit accurate homologous pairing, but banding patterns can reveal chromosomal defects impossible to discern by morphological characterization alone.

Microassays. One of the largest problems of prenatal diagnosis involves the relatively large volume of cells required for biochemical assay and the time required to grow them. As long as six weeks may be needed to culture sufficient amounts of cells for the types of biochemical assays currently available. The development of microassays for the

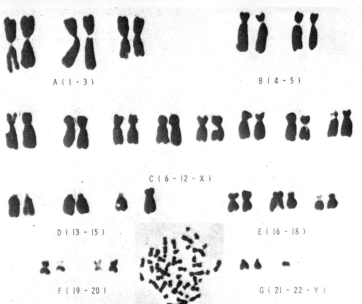

A (1 - 3)   B (4 - 5)

C (6 - 12 - X)

D (13 - 15)   E (16 - 18)

F (19 - 20)   G (21 - 22 - Y)

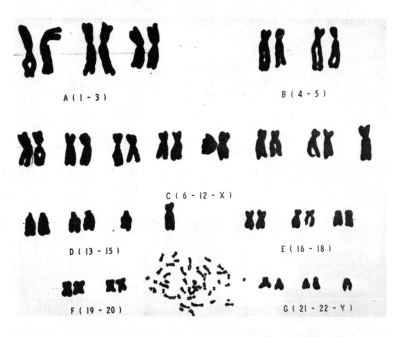

Dr. Henry Nadler confers with a husband and wife considering parenthood (top). The Northwestern University pediatrician explains that the mother is a translocation carrier whose children are likely to have Down's syndrome. Her chromosome karyotype (middle) indicates an abnormality in the No. 21 chromosome. A karyotype of the fetus (bottom) indicates that Down's syndrome will result from the extra No. 21 chromosome. (NIH photos.)

various enzymes involved would solve this problem to a great extent. Dr. Nadler is encouraged by the development of histochemical techniques sensitive enough to permit analysis of significantly fewer cells than are now required. If these techniques prove accurate, as few as five cells taking only three to four days to grow could be assayed.

Although analysis of uncultured amniotic cells is possible and has been performed with a degree of success, most work to date has been done with cultured cells. Researchers indicate that the reliability of a diagnosis on the basis of uncultured cells is questionable and should be avoided. Some cells harvested from the amniotic fluid are dead and can therefore lead to an erroneous diagnosis.

Uncultured cells are most frequently used to determine sex of the fetus as early as possible to facilitate the management of pregnancies in women heterozygous for X-linked disorders such as hemophilia. Amniotic fluid cells are harvested from cultures following the first day of incubation and are stained with fluorescent dyes for sex chromatin. This staining technique reveals the presence of a Y chromosome, indicating that the fetus is male. However, in cases where no fluorescence is present, conclusions based upon this technique alone are equivocal. The Y chromosome may have failed to fluoresce for a variety of reasons, the fetus may be female or only maternal cells may be present in the fluid. Furthermore, it is unlikely though possible that an X-linked disorder such as Turner's syndrome may interfere with a correct analysis of sex determination. For these reasons, researchers such as Dr. Macintyre recommend that sex chromatin analysis of uncultured cells be confirmed through karyotypes of cultured cells before diagnosis is finalized.

A problem area of particular concern to many clinicians results in cases of multiple pregnancies. In these situations there is a distinct possibility that amniotic fluid will be inadvertently sampled from only one of the amniotic sacs and the resultant cell cultures will therefore represent only one fetus. The development of ultrasonic techniques has recently made detection of multiple pregnancies possible early in gestation. As early as the eighth week of pregnancy ultrasonic probes can scan the surface of the abdomen and bounce back a series of echoes which can then be interpreted to identify and localize the fetuses.

Sound waves pass through abdominal tissue at variable speeds according to density. From the pattern of echoes bounced back and viewed on a television screen, doctors can determine the presence of multiple fetuses and pinpoint their positions in the uterus. Localization of the fetus affords information for a more effective amniocentesis and by alerting doctors to multiple pregnancies indicates the need to draw fluid from each sac and to maintain separate cultures.

Caution. Despite the fact that amniocentesis has come a long way in the last several years—two years ago an estimated 200-300 had been performed in the world; today some 1500 have been successfully carried out in the U.S. alone—most scientists and clinicians agree that the technique still remains rather crude and should not be undertaken by any except those with a reasonable amount of experience and expertise.

Dr. Nadler recommends that a series of requirements be satisfactorily fulfilled before intrauterine diagnostic measures are undertaken. These include an obstetrician experienced in the technique of amniocentesis, laboratory personnel experienced in cultivating amniotic fluid cells and in selecting and developing reliable spreads for karyotypes, technicians experienced in performing the specific diagnostic tests required, and the ability to provide suitable treatment. Treatment generally consists of therapeutic abortion. Dr. Henry Kirkman, Chief of Laboratories at North Carolina Memorial Hospital reports that amniocentesis is not performed on those patients for whom abortion is out of the question for either religious or personal reasons.

Diagnostic amniocenteses are recommended by physicians at the Genetics Referral Center of the University of California at San Diego on the basis of the following statistical portrait: a previous history of hereditary disorders; the mother is a known or suspected carrier of an X-linked disorder; the parents are known carriers of a chromosomal translocation; a previous child was born with a chromosome abnormality; the maternal age at conception was 35 or more years.

Not a total answer. Nor does amniocentesis provide the answers for all genetically related disorders. The two most common genetic defects, sickle cell anemia and cystic fibrosis, cannot as yet be detected prenatally with any degree of certainty. Methods for both of these are on the horizon and it seems only a matter of time before they are perfected.

Dr. Barbara Bowman at the University of Texas has demonstrated that serum from patients afflicted with cystic fibrosis inhibits the action of oyster cilia. The technique has not, however, yielded clear-cut separations between carriers and those afflicted with the disease. When these and other problems have been ironed out, Dr. Bowman hopes to adapt the technique to amniotic fluid for diagnosis of the disease *in utero.*

Prenatal diagnosis of sickle cell anemia presents different problems. According to Dr. Nitowsky, methods for identifying the sickle cell genotype are currently available, but prenatal diagnosis is dependent upon obtaining specimens of fetal blood, something which has never been attempted. With the development of the "fetal amnioscope" drawing of fetal blood may become a possibility.

If small amounts of fetal blood could be drawn, red cells from the fetus which normally manufacture Hemoglobin F could be "tricked" into synthesizing adult hemoglobin by inhibiting production of normal fetal hemoglobin. Once fetal hemoglobin is inhibited, the red cells would be compelled to produce Hemoglobin A and /or Hemoglobin S, indicating a propensity towards the disease. Using chromatography, cells could then be separated and an evaluation made.

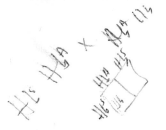

The Therapy of Genetic Diseases
Donald L. Rucknagel, MD

IN RECENT years the hopelessness surrounding genetic diseases has given way to attempts, in some cases relatively successful, at their treatment. In general, successful treatment is based upon an understanding of the biochemical basis of the disease in question and an understanding of the underlying physiologic disturbance. Before discussing the details of treatment methods, however, a discussion of the mechanisms of genetic disease in general terms is in order.

A large array of abnormal human hemoglobins contribute much to our understanding of the phenotypic effects of mutations. Chief among these is sickle cell anemia so I shall use that as a prototype to define gene action.

Sickle cell anemia as a genetic prototype. Sickle cell anemia is due to homozygosity for the gene for hemoglobin S. Affected individuals have a chronic lifelong anemia. Periodically episodes of pain of varying intensity and lasting hours or days may affect various parts of the body. The heterozygous state, referred to as sickle cell trait, is asymptomatic and should not be considered a disease. When persons with sickle cell trait have children by a spouse with normal hemoglobin, the trait is transmitted to half their off-spring. When both parents have the trait the odds that each child will have normal hemoglobin, sickle cell trait, or sickle cell anemia are 25, 50, and 25 percent, respectively. These odds are independent of the outcome of previous pregnancies.

The biochemical basis of the sickling phenomenon is the substitution of valine for the glutamic acid at the sixth position from the amino terminal end of the beta polypeptide chains of the hemoglobin molecule (the hemoglobin molecule is composed of two alpha and two beta polypeptide chains with 141 and 146 residues, respectively). This single amino acid substitution favors aggregation of hemoglobin molecules when the hemoglobin is deoxygenated. As a result the sickled red cell is more fragile and blood viscosity is increased, causing the anemia and vascular occlusion which is responsible for the intermittent episodes of pain characteristic of this disease.

A large number of rarer genetic mutants of hemoglobin alter the physiologic function of the molecule. Most of these are due to single amino acid substitutions at other locations in the molecule causing diminished oxygen affinity, methemoglobin formation, increased oxygen affinity, or molecular

THE SCIENCE TEACHER, 1973, Vol. 40, No. 8, pp. 20-22.

instability resulting in denaturation. The correlation between the structural abnormality and the physiologic dysfunction provide insight not only into the laws governing hemoglobin structural integrity and function but also into the properties of genes, proteins, and enzymes generally. Indeed, a gene is the genetic information that codes the amino acid sequence of a protein polypeptide chain. The proteins encoded by mutant genes differ from normal usually by substitution of single amino acid molecules. When an amino acid substitution interferes with the active site of an enzyme or disturbs the structural integrity of the molecule, the biologic function catalyzed by that enzyme is impaired. Therapy, therefore, is designed to offset the effects of that impairment in the following manners.

Therapy by dietary manipulation. Phenylketonuria is a recessively inherited disease in which the individual is unable to metabolize phenylalanine. Phenylalanine is a so-called essential amino acid which cannot be synthesized by the body. It is either incorporated directly into protein or converted to the amino acid tyrosine which is also a constituent of protein. In phenylketonuria the enzyme activity which catalyzes this conversion, phenylalanine hydroxylase, is deficient. This is inconsequential as far as tyrosine metabolism is concerned, since tyrosine is also present in the diet. Phenylalanine, on the other hand, accumulates in the blood, since it cannot be converted to tyrosine. The large amount of accumulated phenylalanine is then converted to other phenylalanine metabolites, especially phenylpyruvate. Phenylpyruvate, from which the name of the disease is derived, is a ketone which imparts a characteristic odor to the urine of affected individuals.

It is not clear whether the severe mental retardation, the most important feature of the disease, is due to accumulation of phenylalanine, of phenylpyruvate, or of another metabolite. In any event, if the disease is detected soon after birth and the infant placed on a diet containing just enough phenylalanine to assure normal protein synthesis, the accumulation of toxins derived from phenylalanine is prevented. After approximately six years of age, when brain development has largely ceased, phenylalanine can then be ingested with impunity, even though phenylalanine and its metabolites accumulate in the blood again. Because this diet prevents severe mental retardation, routine testing of all newborns for PKU is now mandatory. This therapy does create another problem, however, to which Dr. Neel will return. Namely, women who have been "saved" by the diet bear children who are severely mentally retarded, even though their husbands are normal and the infant a heterozygote at most.

A similar diet therapy is employed in the treatment of galactosemia. In this disease large amounts of galactose accumulate in liver, spleen, brain, and the lens of the eye causing cataracts. The defect is an inherited deficiency of galactose-1-phosphate uridyl transferase activity, either because the enzyme is not produced or is inactivated by a structural abnormality. This enzyme catalyzes the transfer of uridine and phosphate from uridyl disphosphoglucose to galactose-1-phosphate as follows:

UDP-glucose + galactose-1-P

$$\xrightarrow{\text{Transferase}} \text{UDP galactose} + \text{glucose-1-P}$$

This step is necessary for the subsequent metabolism of galactose. Substitution of milk by soy protein in the infant period prevents the accumulation of galactose and, thus, the complications of the disease.

Therapy by detoxification. Another strategy for therapy is to augment excretion of accumulated toxins. In a condition known as Wilson's disease copper is deposited in brain, liver, and other tissues because a copper binding protein, ceruloplasmin, is deficient in the plasma. Penicillamine chelates the copper, allowing it to be excreted in the urine, thus preventing its accumulation. Penicillamine is also employed in the treatment of cystinuria, a recessively inherited disease in which the rather insoluble amino acid cystine precipitates in the kidney. In this disease the basic defect is impairment of transport of cystine and other amino acids across renal tubular cell membranes. Penicillamine forms soluble mixed disulfides with cysteine, preventing the precipitation of cystine in the kidney and urinary tract. In gout, which is due to both environmental and hereditary factors, uric acid excretion is favored by allopurinol, a substance which blocks the formation of uric acid, allowing its excretion in a more soluble form, hypoxanthine.

Therapy by administration of deficient proteins. Another approach is to substitute the missing gene product. One well-known example is hemophilia, in which either normal plasma containing the antihemophilic globulin or the purified AHG are administered to stop hemorrhaging. Also in diabetes, which has a more complex genetic component, insulin is injected. This mode of therapy depends upon the availability of the protein and is not practical for most enzyme deficiencies.

Cofactor therapy. In some diseases therapy may be directed toward the variant enzyme per se. In these instances the structural abnormality produced by the gene has presumably altered the enzyme's ability to bind a vitamin "coenzyme" necessary for the metabolic reaction catalyzed by that enzyme. Administration of large amounts of the vitamin cofactor allows the reaction to proceed. For example, an inherited block in the metabolism of methylmalonic acid responds to administration of vitamin B_{12}. Also there is an anemia that is inherited in a sex-linked fashion and that mimics iron deficiency. It responds to pyridoxine or vitamin B_6 therapy.

Surgical therapy. Even the surgeon may be called upon to treat genetic diseases. For instance, cleft lip and palate are believed to be inherited as polygenic traits. Pyloric stenosis is due to a congenital hypertrophy of the pyloric sphincter, the muscle regulating the flow of stomach contents into the small intestine. This condition is inherited as a polygenic trait. The obstruction is relieved by excising a portion of the enlarged muscle, thus allowing emptying of the stomach. More recently, renal transplants are being attempted to treat Fabry's disease, in which a glycolipid accumulates in the kidney as a result of an alpha galactosidase enzyme deficiency.

"Genetic engineering." Much is being said today regarding the possibility of genetic engineering or genetic surgery, defined as the ability to correct or replace the defective gene itself. A model for such therapy is provided by transduction of bacterial genes by infection with bacteriophage, whereby the phage virus genome becomes incor-

porated for higher organisms into the bacterial chromosome. At the moment this seems to be a long way off, since the mammalian cell nucleus is far more complex than the bacterial chromosome. Moreover, even if a transducing virus could be identified which could infect the proper cell, it could also potentially produce more disease than it cured. Although the barriers to such therapy seem formidable, another form of therapy that is not altogether outlandish is the possibility of switching on fetal genes. For numerous proteins there are distinct fetal forms which then are replaced by an adult type in the process of development. A genetic abnormality of the adult form could be treated effectively if the genes governing the fetal form could remain active during adulthood. For instance, sickle cell anemia might be cured if fetal hemoglobin, the predominant hemoglobin of term fetus red blood cells, could be increased significantly in the red cells of affected individuals.

The treatment of genetic diseases is still in its infancy, and in many instances it is complicated and falls short of a cure. Nevertheless, it is possible to effect significant improvements in the lives of many affected persons, and one can expect further developments and refinements with the passage of time.

Reference

Harris, H. *The Principles of Human Biochemical Genetics.* American Elsevier Co., New York. 1971. P. 328.

GENETIC ENGINEERING: FUTURE PROSPECTS

GENETIC MANIPULATION AND MAN

Darrel S. English

THE SCIENCE OF IMPROVING HUMAN BEINGS by applying the principles of inheritance to obtain a desirable combination of physical characteristics and mental traits is called eugenics. The term was coined by Francis Galton in 1883; literally translated, it means to be "true born" or "well born."

Although most writers on this subject begin by citing the Greeks, the idea of improving the human stock probably goes back even farther. Even though he lacked any knowledge of the laws of heredity, primitive man could see that parents with imperfections often bore children with the same deficiencies. Perhaps the earliest aim was to produce a race of physically perfect men, capable of coping most efficiently with the tremendous hardships they had to endure, including contests with enemies and wild beasts. Paralleling the desire to develop physically was the need to develop intellectually; and cultures may have tended to develop their mental capacities more than their physiques.

Plato advanced the idea of race improvement by methods similar to those of present-day stock breeders. In *The Republic* he proposed that matings between the most nearly perfect men and women be encouraged and that their offspring be raised in a state nursery. Inferior persons should be prevented from reproducing; and if by chance they should have children these should be destroyed. To some extent Plato's eugenic methods were practiced in Sparta; the result was a population of people with fine physiques (Castle, 1925; Fasten, 1935).

In Athens the emphasis was on art, politics, and science. Here, too, people of good background were encouraged to marry among their kind. Did this pay

THE AMERICAN BIOLOGY TEACHER, I972, Vol. 34, pp. 507-513, 526.

off? Galton (1909) noted that during the 6th to 4th centuries B.C. Athens produced some of the most illustrious men the world has ever known. Whether the decline of Greek society was due to master–slave intermarriage, as some have suggested, or whether the "inferior" classes reproduced more rapidly than the "superior" classes, one can only guess. Possibly the downfall resulted from economic rather than genetic changes.

The eugenic movement appears to have made little headway until the 19th century. Charles Darwin, Francis Galton, and Gregor Mendel were among the scientists who kindled the spark of modern eugenics; directly or indirectly, their work stimulated interest in this field. The Darwinian concept of the "survival of the fittest" brought to the fore many inescapable implications. In *The Descent of Man* (1874) Darwin wrote: "It is our natural prejudice and arrogance which made our forefathers declare that they were descended from demi-gods and which leads us to demure to this conclusion." The growing evidence supporting the theory of evolution, together with the refinement of man's ability to influence the evolution of domesticated plants and animals, stimulated work along eugenic lines.

Darwin's cousin Francis Galton, the English anthropometrist and examiner of family records, led the first big surge toward eugenic studies. In his book *Enquiries Into Human Faculty and Its Development* (1883) he coined the term eugenics, defining it as the study of agencies, under social control, that could improve or impair the hereditary qualities of future generations, either physically or mentally (to paraphrase the 2nd ed., 1908). He proposed improvement in human breeding by decreasing the birthrate of unfit persons and increasing the birthrate of fit persons. He made extensive studies on criminality, insanity, blindness, and other human defects. Galton was able to understand the inheritance pattern of some human traits. He recognized the importance of twin studies for human genetics and was aware of the social implications of genetic change in man. He was instrumental in applying more sophisticated statistical methods of solving problems of genetics. (It is interesting to note that Galton, who was unusually gifted and was devoted to the principle that better-qualified people should produce at least their share of children, himself died childless.)

Mendel shed new light on the genetics of man with

the discovery of fundamental principles of inheritance. Thanks in part to his work, scientists learned how to deal with questions of human heredity in a methodical manner.

Eugenic Implementation

Eugenics is often divided into "negative," or preventive, and "positive," or progressive, eugenics. The first is concerned with the elimination of alleles that produce undesirable phenotypes; the second is concerned with furthering the increase of alleles that produce desirable phenotypes or at least with guarding against the decrease of these alleles. In a sense, the two branches of eugenics are identical: to discourage reproduction among people having undesirable traits is, ipso facto, to encourage a comparatively higher rate of reproduction among people who lack these undesirable traits.

In the past, negative eugenics meant sterilization and institutionalization. The results of these measures were insignificant. In many cases, determining what should be considered undesirable was a major problem; furthermore, procedures were not always ethically acceptable to the majority of the people.

More recent methods of negative eugenics include dissuasion from procreation, voluntary sterilization, medically induced abortions, education as to the genetic basis of human traits, and the encouragement of birth-control practices by persons possessing undesirable traits. These efforts are intended to reduce the dysgenic, or deteriorating, effect on society caused by the perpetuation of certain undesirable traits.

Some biochemical defects, which are known to be propagated by a single defective gene, tend to eliminate themselves naturally; many, however, do not. Hemophilia, for example, has severe and often lethal effects generation after generation. In agammaglobulinemia, children are born without the ability to manufacture antibodies. In phenylketonuria (PKU), children are unable to metabolize phenylalanine; they become mentally incompetent if not treated soon after birth. In certain conditions the genetic defect may not be discovered until after reproductive age has been reached and children have already been introduced into the population. Huntington's chorea, with its progressive deterioration of the muscular and nervous systems, does not make itself known until the victim is in his forties. As a result of this dominant lethal gene, approximately 50% of the offspring can expect to succumb to the same fate.

Positive eugenics, aiming at the reproduction of persons of presumedly superior genotypes, has as many problems associated with it as negative eugenics. This concept came into disrepute because of early notions of who was "desirable" and the classification of certain kinds of people as "degenerates." A major setback for eugenicists came during World War II, when Hitler's eugenic movement went to the extreme of trying to achieve a "master race" of "Aryans" at the expense of "non-Aryans." It is understandable why the term eugenics has some very bad connotations in the minds of many people.

Unfortunately, many characteristics that we might consider desirable—high intelligence, good physical health, longevity—are not under the control of a single genetic factor; instead, they arise from a complex of genes that interact in an appropriate environment. H. J. Muller, J. F. Crow, and others have pointed out that such traits as high intelligence and esthetic sensibility have not been selected with any effort in the past. They postulate, however, that if selection for such traits were to be instituted, the general population might respond very rapidly. The means of selection remains the paramount problem. Selection schemes, even though they might be extremely successful, may prove to be completely intolerable.

During the early 1960s, Muller hotly advocated the use of sperm banks in preference to selective-mating programs. These banks would preserve, frozen, the sperm of men of outstanding qualities. This method, called germinal choice, or eutelegenesis, assumes that married women, otherwise barren, might choose to be artificially impregnated with semen from men who had highly desirable traits. This would be possible even though the donor had died several years before conception took place. Full information about the donor would be provided to ensure the best possible combination of genes (Carlson, 1972).

The possibility of such a "preadoption" method has had some acceptance in the United States. It is estimated that 10,000 artificially inseminated conceptions occur in the United States every year (Taylor, 1968). The reason for most births of this kind is that the husband is impotent or possesses some genetic incompatibility, such as the Rh factor, or harbors a known genetic defect, such as hemophilia.

Euphenics

Molecular biologists and medical researchers are developing a series of procedures for the relief of genetic disorders. The field of euphenics is concerned with the improvement of genotypic maladjustments by treatment of genetically defective persons at some time in their life cycle. Today there are many sensitive tests for genetic defects. These enhance the reliability of counseling and decision-making before or during childbearing. As for treatment: in phenylketoneuria, for example, the child is given the "diaper test," which depends on a color change of the urine when ferric chloride is added; or the Guthrie test, which is based on the ability of certain strains of bacteria to grow on substrates containing high levels of phenylalanine. If a problem exists in the metabolism of phenylalanine and the condition is diagnosed early, the infant is put on a diet low in phenylalanine for the first five years of life, and the brain develops normally.

Another example of technologic success in the detection of genetic abnormality is that of a woman who possessed the potential to produce a mongoloid child. The woman had three sisters who suffered from Down's syndrome, or mongolism, and she feared that her own children might have this condition. In 1959 it was discovered that the syndrome is due to an additional (21st) chromosome, which originates by a mistake in cell division just before conception. The woman requested a study of her cells, and her fears were confirmed: she had the extra chromosome. Genetic counselors advised her that she had one chance in three of producing an abnormal child. Several years ago the woman became pregnant. Doctors informed her that a new technique, called amniocentesis, would enable them to determine whether the fetus was aberrant. The method consists in tapping the fluid of the amnionic sac and making chromosome studies of the cells from the fluid. During the 14th week of pregnancy the woman's fears were borne out: she was told her baby was mongoloid. The pregnancy was terminated by therapeutic abortion. Several months later she became pregnant a second time; once again the tests showed Down's syndrome, and the pregnancy was terminated. There was a third pregnancy; and this time the chromosome studies indicated the baby would be normal—and a boy. The woman at last gave birth to what she had wanted for so long: a normal son. A year and a half later, following careful testing, she gave birth to a normal daughter.

Henry Nadler has used amniocentesis in the diagnosis of high-risk mongolism cases with a high degree of accuracy. This test can be performed between the 12th and 18th weeks of pregnancy. Although there is some danger, the benefits are said to more than justify the risks. Over 35 human enzymatic diseases have been identified by this technique; they include cystic fibrosis, cystinosis, amaurotic idiocy, gout, Gaucher's disease, galactosemia, xeroderma pigmentosum, and diabetes mellitus. Recent advances in the detection of carriers of recessive diseases, such as hemophilia and some forms of muscular dystrophy, are helping to make the job of genetic counseling an easier one (Friedmann, 1971; Nadler and Gerbie, 1971; *Time*, 1971).

Human genetic analysis of single-gene effects by pedigree analysis is still valid and useful. McKusick (1970) listed 1,487 human traits known to be controlled by a single dominant gene, 531 by a recessive gene, and 119 by X-linked genes. With the advent of computerized technology, experiments may now be designed for determining the genetic mechanisms from family-history data, and additional information will be rapidly added to the catalogue of human genetic defects.

From the viewpoint of the population geneticist, the symptoms of the hereditary diseases may have been treated, but the genes remain unchanged and can be passed on to subsequent generations. Therefore the real genetic problem is not solved and, in fact, such medical practices only compound future problems: by preserving defective genotypes and allowing them to reproduce and transmit these genes, we create a population that is more dependent upon surgery, drugs, and similar treatments.

Furthermore, to simply prevent people who possess defective genes from reproducing will not solve the problem. It is estimated that, on average, each person carries four to eight defective genes that in combination with other defective alleles could bereft a child of his normal faculties. In other words, each conception involves some risk of producing a child with a serious abnormality.

The eugenics procedures of the future may be quite different from those of the past. It now appears that the techniques may be at hand to not only "improve" the genetic material by selection but also to correct misinformation within the DNA molecule and thus eliminate the problem.

Genetic engineering, or algeny, is becoming a household word among molecular biologists. Terms such as gene surgery, gene insertion, and gene deletion are beginning to have real meaning for the future of man. The ability to manipulate, in a purposeful manner, the genetic constitution of human beings may come sooner than we anticipate. Taylor (1968) asserted that biologists have reached the critical point of sudden acceleration—the point that physicists reached only a generation ago. It is hoped that biologists and laymen alike will face the problems and potentials of human reproductive engineering in a more realistic and relevant manner than was taken by the atomic physicist and the politicians at the creation of the world's most destructive force (Heim, 1972).

Algeny is a more sophisticated and direct method of curing genetic ills. It is also a permanent cure: in certain instances it might alleviate suffering in future generations. Two approaches seem possible:

1. One might consider the incorporation of a normal gene within the protein coat of some human virus. The cells, after infection with the virus, would supply the defective cell with a corrected copy of the needed information. Although this would only cure the individual during his lifetime and not have any effect on his progeny (unless there might happen to be accidental gene-incorporation), it would be a more direct attack on the genetic problem.

This approach does not appear to be out of the question—as was accidentally demonstrated in a group of laboratory technicians and doctors working with Shope papilloma virus. This virus is capable of inducing tumors in rabbits. Although nonpathogenic in humans, it does infect those who handle it. It appears that one of the viral genes, which is responsible for the synthesis of arginase, is active in human cells. Years after any contact with the laboratory, infected workers were shown to have especially low levels of arginine in their blood as a result of the activity of the viral enzyme (Taylor, 1968).

And now genetic engineers are making their first attempts at using a related means of treating this metabolic disease. Investigators at the Oak Ridge National Laboratories and in Cologne are cooperating in the treatment of two German girls who are suffering from low levels of arginase in their blood. It is hoped that injections of live Shope papilloma virus carrying the enzyme will supplement their low levels of arginase. If this is successful, the girls will be able to produce the needed amino acid, arginine, and so achieve a more nearly normal metabolic balance. Someday it may even be possible to produce viruses artificially to correct specific metabolic deficiencies (Gardner, 1972).

2. A second approach to genetic engineering involves replacement of a mutant gene with a normal one by treating germ cells before fertilization. This curative procedure would be even more direct and would have the advantage of being permanent, because descendants would be normal with respect to the defect in question.

In the microbiologic world two methods of incorporating normal genetic material into a mutant cell have been perfected. Avery, MacCleod, and McCarty (1944) exposed nonvirulent bacteria to purified DNA of a virulent strain of bacteria. To their amazement the nonvirulent bacteria were transformed into virulent forms and were able to transmit the trait to future generations. This process is called transformation.

In 1959 three French workers reported that they had extracted DNA from one strain of ducks, called Campbells, and injected the material into a second strain, Pekins. They had hoped to change the offspring somehow, but to their astonishment the injected Pekin ducks began to take on some of the characteristics of the Campbells. This stimulated a flurry of experiments with these strains of ducks, as well as other animals. But, to the dismay of the researchers, subsequent attempts were a total failure, even with the ducks. Not until 1966 was an actual case of tranformation verified in organisms other than bacteria. A. S. Fox and S. B. Yoons, of the University of Wisconsin, treated one strain of fruit flies with DNA extracts of another strain; some of the resulting offspring developed genetic anomalies that persisted for several generations (Taylor, 1968).

In contrast with transformation is another mode of permanent genetic change effected by using viruses.

This process, called transduction, involves a viral particle that picks up a host gene and, on reinfection in a second host, gives up the genetic material to its new host genome. After incorporation of the newly introduced genome the cell is permanently altered and transmits the acquired trait in the typical genetic manner. The Ukrainian scientist Serge Gerhenson claimed to have transduced a silkworm, using a virus to introduce the foreign DNA. Likewise, there are lines of evidence suggesting that other investigators have been able to transfer drug resistance from one line of mouse-cell cultures to another in this manner (Taylor, 1968).

More recently an exciting experiment was carried out by the molecular biologist Carl Merril and his colleagues the first successful transplant of bacterial genes into living human cells. Cells from a victim of the genetic disease galactosemia were cultured in vitro. These cells are unable to produce an essential enzyme for the breakdown of the simple sugar galactose. Newborn infants with this defect face malnutrition, mental retardation, and death unless they quickly receive a milk-free diet. Using viruses that had picked up the genes for galactose metabolism from the common intestinal bacterium *Escherichia coli,* the researchers hoped to transmit the gene to human cells in tissue culture. Further investigations showed that the cells had picked up the viruses and that they were producing messenger RNA for the missing enzyme and the enzymes themselves. This clearly implied that the cells were being directed to produce the essential enzyme; and equally exciting was the fact that the enzyme-making capabilities were being transmitted to future generations of cells (Merril, Geier, and Petricciani, 1971). These investigators are now striving to make the same kind of genetic transplants with laboratory animals.

These results are particularly significant because they show that bacterial genes can become biologically active in mammalian cells. Furthermore, they clearly establish the universality of the genetic code. And they could have some important implications for the cure of cancerous conditions produced by "runaway" genes. The field of genetic engineering is indeed wide open. It is enough to make one wonder, though, what new genes a person may pick up from the sneeze of the person sitting next to him!

Clonal Reproduction

Once a highly desirable genotype has been produced, it would be beneficial if more of the same organism could be produced. Asexual reproduction of organisms, including man, is another method that may be used someday to produce desirable genotypes. Cloning is the process of inducing normal somatic cells to repeat the complex step in embryogenesis and eventually produce carbon copies of the original donor organism.

F. C. Steward first showed the feasibility of cloning by taking certain cells from a carrot root and culturing them in coconut milk. Some of these cells formed clumps, which began to differentiate. Transferred to soil, they matured into normal carrot plants. Later experiments have shown that almost any early embryonic cell of the carrot can grow vegetatively (Steward, Mapes, and Smith, 1958).

The possibilities of animal cloning took on reality when J. B. Gurdon, of Oxford University, managed to get the nucleus of an intestinal cell of the South African clawed toad, *Xenopus laevis*, to direct embryogenesis in the enucleated cytoplasm of an unfertilized egg cell. The egg, thus, contained the diploid set of chromosomes and responded by dividing repeatedly. The resulting tadpole was a genetic twin of the toad that had provided the nucleus. By making numerous subclones Gurdon was able to produce many identical copies of the parent toad (Gurdon, 1968). More recently he was able to prepare frogs from cultured cell nuclei (Gurdon, 1970). It would seem there might be no end to the number of copies possible from a single individual. These experiments prove that all the genetic information necessary to produce an organism is encoded in the nucleus of every cell of the organism.

The implications of this technique are numerous. It might be possible to clone a group of Einsteins or Beethovens. Or one might desire a team of astronauts with particular talents and temperaments for a long space voyage. One might be able to clone people with acute psychic awareness, so that extrasensory perception would become commonplace.

A further advantage of this technique has to do with immunologic properties: the members of a clone would be able to accept grafts and tissue or organ transplants without any of the usual repercussions. (The recipient may, however, have some difficulty in convincing one of his clonal twins to give up his heart, lungs, kidneys, or limbs!)

A more practical use of cloning would be its use as a means of tracking extremely deleterious genes during early embryologic development. Someday it may be possible to take the fertilized egg or embryo and culture it in a test-tube. A few cells could be removed and cultured in sufficient numbers for biochemical analysis. If the embryo proved to possess the deleterious genotype, the culture could be terminated; if not, the egg or embryo could be reimplanted in the womb, where development would proceed normally and without further interruption.

Taylor (1968) noted that if vegetative reproduction of human beings is ever achieved, it is most likely to be done by growing a few cells taken from the embryo. The more specialized a cell becomes, the greater the loss of its totipotency. To induce specialized cells, such as nerve, muscle, or brain cells, to become unspecialized once again, appears to be an extremely difficult task.

Of more immediate use to the economy of the world is what one might call a variation on the theme of cloning: the phenomenon of artificial inovulation. All animals seem to be capable of producing many more eggs than they will ever release normally. Injections of the follicle-stimulating hormone (FSH) can induce as many as 40 or more eggs at one time in a cow. This is called superovulation. If the eggs are then artificially fertilized and implanted in a number of competent females, the number of offspring can be multiplied manyfold annually. This process is the converse of artificial insemination, which enables a prize bull to sire more than 50,000 calves a year.

In 1962 two South African ewes gave birth to two lambs whose real parents lived in England. The fertilized eggs had been implanted in the oviduct of a live rabbit, which was flown to South Africa; there the eggs were implanted in the foster ewes. Since then, eggs have been flown to the United States successfully, and transfers between different strains of animals have been accomplished (Taylor, 1968).

Experiments of this kind promise valuable improvements of livestock worldwide. Artificial inovulation also offers hope to those humans afflicted with certain types of sterility. Perhaps the day will come when a woman who would normally be childless will have a prenatally adopted child implanted in her womb, and she will be able to experience all of the emotions and physiologic changes associated with motherhood.

Also related to the ability of man to manipulate the activity of a cell is the phenomenon of regeneration. Mammals have generally lost their totipotency except in specific parts, such as the liver, lymphoid tissue, skin, and bones. The initiatory and regulative factors of regenerative growth are generally unknown. If, however, one could reactivate the genetic events involved in the embryonic organization of cells at the stump of a lost limb, might it not be possible to regenerate the entire structure?

Robert O. Becker, of the Veterans Administration Hospital in Syracuse, N.Y., has said that the possibility of regenerating limbs of mammals, including man, seems nearer. He has successfully stimulated partial regeneration of limbs in rats by the induction of a blastema in response to minute electrical currents applied to the severed area (Becker, 1972). The production of several centimeters of regenerative growth in experimental mammals suggests that higher animals do have regenerative potential if the cells involved can be induced to take on the more primitive state of development. Eventual total regeneration of organs and limbs would make the immunologic complications in transplants a thing of the past.

Conclusion

At one time artificial insemination in humans, sex determination before birth, sex reversal, embryonic determination of genetic aberrations, and organ transplants were looked upon as remote possibilities. Today they are facts; and society has, to some degree, accepted them.

Will man be able to alter his own makeup so profoundly that he will be essentially a new species? Undoubtedly he will strive, with even more enthusiasm, to unlock more of nature's secrets. But some scientists see far greater dangers in man's new knowledge than may be envisioned at first. Seymour Kessler, of Stanford University, has said he "would hate to see manipulation of genes for behavioral ends because as man's environment changes and as man changes his environment, it is important to maintain flexibility" (quoted in Taylor, 1968). One must be cautious about following a path that eliminates variability, because without it we could go the way

of the dinosaurs. This is a particular danger if cloning should become popular. Thus, with increased
knowledge comes the responsibility to use this new
information wisely (Heim, 1972).

Many biologists are concerned over the moral,
ethical, and social implications of the new biology.
Although predictions of human genetic control are
probably premature—many of the techniques are far
from perfected—it is only a matter of time before
test-tube babies, cloning of humans, and corrective
gene-surgery will be possible.

Numerous national and regional commissions have
been set up to study some of the problems. Joshua
Lederberg, of Stanford University, does not believe
that the perfect human being is a proper goal for the
molecular geneticist, even if the techniques could be
perfected. Nevertheless, there are those who disagree. Who, then, decides what qualities are to be
preserved and by whose standards are they to be
directed? J. D. Watson, in an article entitled, "Mov-
ing toward the Clonal Man: Is This What You
Want?" said that he hopes that mankind will
thoroughly discuss these issues in the next decade.
On the other hand, Glenn T. Seaborg, former director of the Atomic Energy Commission, has stated
that he believes decisions should be made by experts
without the benefit of public debate. The latter comment reminds us of the warning of the late C. S.
Lewis, over a quarter of a century ago: "Man's power
over Nature is really the power of some men over
other men, with Nature as their instrument." One
may agree with Senator Walter F. Mondale, who
said: "There may still be time to establish some
ground rules." And one may doubt whether we really
need a clone of people or the creation of a laboratory
full of orphaned test-tube babies at a time when the
population explosion is one of man's greatest biologic
problems. (For citations of views in this paragraph
see Taylor, 1968, and Wallace, 1972.)

And yet . . . genetic engineering does promise to
alleviate many of man's ills and sufferings. What a
relief from anxiety it would be to know that a genetic
disease you might harbor would not be passed on to
your child! Can one begin to express the gratitude
of a mother as she watches her child, who might have
been mentally retarded a few years ago, present the
valedictory speech at her high-school graduation?
Amniocentesis is likely to show enormous cost benefits in the treatment of Down's syndrome in the near

future: 4,000 mongoloid infants are born each year in the U.S.; the lifetime institutional care for each one is approximately $250,000; and therefore, unless women carrying such abnormal fetuses are encouraged to have therapeutic abortions, their care will cost society some $1.75 billion nationally by 1975 (Friedmann, 1971; *Time*, 1971).

One can see that man faces many new legal, moral, and ethical questions. Will he have to consider assigning certain rights to the fetus? At what point does a fetus have legal rights, and to whom does the test-tube baby turn for support and legal inheritance? What will be the nature of the conflict in our society between personal choice and governmental control? These are serious questions, for which today there are no satisfactory answers.

REFERENCES

AVERY, O. T., C. M. MacCLEOD, and M. McCARTY. 1944. Studies on the chemical nature of the substance inducing transformation in pneumococcal types. *Journal of Experimental Medicine* 79: 137-158.

BECKER, R. O. 1972. Stimulation of partial limb regeneration in mammals. *Nature* 235: 109-111.

CARLSON, E. A. 1972. H. J. Muller. *Genetics* 70: 1-30.

DARWIN, C. 1874. *The descent of man*, 2nd ed. D. Appleton Co., New York.

FASTEN, N. 1935. *Principles of genetics and eugenics*. Ginn & Co., New York.

FRIEDMANN, T. 1971. Prenatal diagnosis of genetic diseases. *Scientific American* 225 (5): 34-51.

GALTON, F. 1908. *Inquiries into Human Faculty and Its Development*, 2nd ed. E. P. Dutton Co., New York.

————. 1909. *Essays on eugenics*. Eugenics Education Society, London.

GARDNER, E. J. 1972. *Principles of genetics*, 4th ed. John Wiley & Sons, Inc., New York.

GURDON, J. B. 1968. Transplanted nuclei and cell differentiation. *Scientific American* 219 (6): 24-36.

————. 1970. The transplantation of nuclei from single cultured cells into enucleated frog's eggs. *Journal of Embryological and Experimental Morphology* 24: 227-248.

HEIM, W. 1972. Moral and legal decisions in reproductive and genetic engineering. *American Biology Teacher* 34 (6): 315-318.

McKUSICK, V. A. 1968. *Mendelian inheritance in man: catalogs of autosomal dominants, autosomal recessives and X-linked phenotypes*, 2nd ed. Johns Hopkins Press, Baltimore.

MERRIL, C. R., M. R. GEIER, and J. C. PETRICCIANI. 1971. Bacterial versus gene expression in human cells. *Nature* 233:

398-400.

NADLER, H. L., and A. GERBIE. 1971. Present status of amniocentesis in intrauterine diagnosis of genetic defects. *Obstetrics and Gynecology* 38: 789-799.

STEWARD, F. C., M. O. MAPES, and J. SMITH. 1958. Growth and
organized development of cultured cells, I: growth and
division of freely suspended cells. *American Journal of
Botany* 45: 693-703.

TAYLOR, G. R. 1968. *The biological time bomb.* World Publishing Co., New York.

TIME [magazine]. 1971. Man into superman: the promise and
perils of the new genetics. April 19: 33-52.

WALLACE, B., ed. 1972. *Essays in social biology, vol. 2.* Pren-
tice-Hall, Inc., Englewood Cliffs, N.J.

Gene Therapy for Human Genetic Disease?

Theodore Friedmann and Richard Roblin

At least 1500 distinguishable human diseases are already known to be genetically determined (*1*), and new examples are being reported every year. Many human genetic diseases are rare. For example, the incidence of phenylketonuria is about one per 18,000 live births or about 200 to 300 cases per year in the United States (*2*). Others, such as cystic fibrosis of the pancreas, occur about once in every 2500 live births (*3*). When considered together as a group, however, genetic diseases of humans are becoming an increasingly visible and significant medical problem, at least in the developed countries. While the molecular basis for most of these diseases is not yet understood, a recent review (*4*) listed 92 human disorders for which a genetically determined specific enzyme deficiency has been identified.

Concurrent with the recent progress toward biochemical characterization of human genetic diseases have been the dramatic advances in our understanding of the structure and function of the genetic material, DNA, and our ability to manipulate it in the test tube. Within the last 3 years, both the isolation of a piece of DNA containing a specific group of bacterial genes (*5*) and the complete chemical synthesis of the gene for yeast alanine transfer RNA (*6*) have been reported. These advances have led to proposals (*7*) that exogenous "good" DNA be used to replace the defective DNA in those who suffer from genetic defects. In fact, a first attempt to treat patients suffering from a human genetic disease with foreign DNA has already been made (*8*).

Nevertheless, we believe that examination of the current possibilities for DNA-mediated genetic change in humans in the light of some of the requirements for an ethically acceptable medical treatment raises difficult questions. In order to focus the discussion, in this article we concentrate on the prospects for using isolated DNA segments or mammalian viruses as vectors in gene therapy. For this reason we do not discuss other techniques, such as cell hybridization (*9*), which have been used to introduce new genetic material into mammalian cells. We limit our discussion to the possible therapeutic uses of genetic engineering in humans. The potential eugenic uses, for example, the improvement of human intelligence or other traits, are not discussed because they will be very much more difficult to accomplish (*10*) and raise rather different ethical questions. Whether genetic engineering techniques can be developed for therapeutic purposes in human patients without leading to eugenic uses is an important question, but lies mostly beyond the scope of this article.

SCIENCE, 1972, Vol. 175, pp. 949-955. Copyright 1972 by the American Association for the Advancement of Science.

Schematic Model of Genetic Disease

Some aspects of a hypothetical human genetic disease in which an enzyme is defective are shown in Fig. 1. The consequences of a gene mutation which renders enzyme E_3 defective could be (i) failure to synthesize required compounds D and F; (ii) accumulation of abnormally high concentrations of compound C and its further metabolites by other biochemical pathways; (iii) failure to regulate properly the activity of enzyme E_1, because of loss of the normal feedback inhibitor, compound F; and (iv) failure of a regulatory step in a linked pathway because of absence of compounds D or F, as in the increased synthesis of ketosteroids in the adrenogenital syndrome (11). In some cases of human genetic disease, accumulation of high concentrations of compound C and its metabolites appears to do the damage. Often a consequence is mental retardation.

The pathway in Fig. 1 is typical of some recessively inherited genetic defects which result in a deficiency of some gene product, usually an enzyme or hormone. In theory, such defects might be corrected by gene therapy, since such techniques might be able to restore the deficient gene product. Other kinds of genetic defects, including those such as the Marfan syndrome which show dominant inheritance and those such as Mongolism that are caused by chromosome abnormalities, could probably not be ameliorated by the kind of gene therapy we emphasize here.

Current Therapy

Human genetic diseases are usually treated by dietary therapy (12), drug therapy, or gene product replacement therapy (11). For example, diets low in lactose or phenylalanine are used as treatments for individuals with galactosemia and phenylketonuria, respectively. Such diets have proved exceedingly effective in galactosemia and have produced a marked reduction in the incidence of mental retardation associated with phenylketonuria. In terms of Fig. 1 this therapy corresponds to restricting the intake of compound A, thus minimizing the accumulation of compound C whose further normal metabolism is blocked.

Drug therapy has been used to block or reduce the accumulation of undesired and possibly harmful metabolites. One example is the inhibition of the enzyme xanthine oxidase with the drug allopurinol to reduce the accumulation of uric acid associated with gout and the Lesch-Nyhan syndrome (13). At present, this method of treatment has been applied to only a few human genetic diseases. Its more general application clearly depends upon the availability of drugs which act selectively on specific enzymes. In another form of drug therapy, drugs which combine specifically with the accumulated compound C are used. An example is the use of D-penicillamine to promote excretion of excess cystine in patients with cystinuria (11).

In theory, some human genetic diseases might be alleviated by supplying directly the deficient enzyme (E_3 in Fig. 1). Recently, attempts to treat Fabry's disease (14), metachromatic leukodystrophy (15), and type 2 glycogenosis (16), by administering the missing enzyme have been reported. Since exogenous enzyme molecules are eventually inactivated or excreted from the body, repeated enzyme injections would be required to manage the diseases in this way. In time, the patients would probably respond by forming antibodies

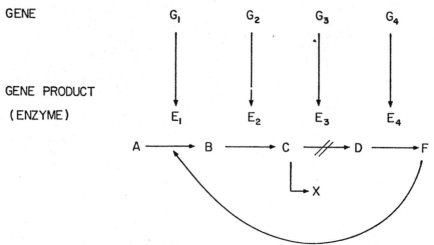

GENE G_1 G_2 G_3 G_4

GENE PRODUCT

(ENZYME) E_1 E_2 E_3 E_4

A → B → C ¬→ D → F

Fig. 1. A hypothetical pathway for the enzymatic conversion of compound A to a final metabolic product F. Compounds B, C, and D are intermediate products. Four different enzymes, E_1, E_2, E_3, and E_4, the products of the corresponding genes G_1, G_2, G_3, and G_4 are required to effect the conversion. The block occurs in the conversion of compound C to D. The concentration of compound F regulates the activity of the first enzyme in the pathway, E_1, in a feedback control loop.

against the administered enzyme. However, insoluble or encapsulated enzyme preparations may in the future provide a means of supplying therapeutic enzymes in a more stable and perhaps less immunogenic form.

There are growing possibilities for the early detection of some genetic diseases by diagnosis in utero. It is now possible to sample the cells of a growing fetus in utero and by examining these cells to diagnose a variety of genetic defects (17). If genetic defects are detected, some states will permit an abortion if the prospective parents so desire. We recognize that diagnosis in utero and abortion raise difficult social and ethical problems of their own and cite them only to indicate that there are additional alternatives to prospective gene therapy for coping with human genetic disease.

However, many genetic diseases do not yet respond to any of these treatments. For example, most genetic disorders of amino acid metabolism (other than phenylketonuria) cannot be well controlled by dietary therapy. Storage diseases associated with lysosomal enzyme deficiencies (18) do not appear to respond to enzyme therapy (14, 15) and will probably be impossible to control by dietary restriction. In addition, even in cases where disease management is effective, it is seldom perfect. Individuals with diabetes mellitus, when treated with insulin, have an increased incidence of vascular and other disorders and a decreased life expectancy compared to the nondiabetic population (19). Children with Lesch-Nyhan syndrome may have their uric acid accumulation controlled by drug (allopurinol) therapy, but their brain dysfunction has, to date, not been reversible.

These limitations of current therapy are stimulating attempts to develop techniques for treating human genetic diseases at the genetic level, the site of the primary defect. Genetic modification of specific characteristics of human cells by means of exogenous DNA seems possible for several reasons. DNA-mediated genetic modification of several different kinds of bacteria has been known for many years (20), and recent experiments suggest that the genetic properties of mutant *Drosophila* strains can be modified by treating their eggs with DNA extracted from other *Drosophila* strains (21). It has also been found that treatment of human cells in vitro with DNA extracted from the oncogenic virus SV40 results in permanent hereditable alteration of several cellular properties (22).

Genetic Modification Mediated by DNA

Permanent, heritable, genetic modification of a human cell by means of DNA requires (i) uptake of the exogenous DNA from the extracellular environment; (ii) survival of at least a portion of the DNA during its intracellular passage to the nucleus; (iii) stabilization of the exogenous DNA in the recipient cell; and (iv) expression of the new genes via transcription into an RNA message (mRNA) and translation of this message into the appropriate protein. Some of these processes are illustrated schematically in Fig. 2.

Mammalian cells take up proteins, nucleic acids, and viruses from their environment by a process known as endocytosis (23). After binding to the cell membrane, the macromolecules are drawn into the cell by an infolding of the external cell membrane leading to vesicle formation (see Fig. 2). Macromolecules contained in vesicles derived from invaginations of the external cell membrane can be degraded if these vesicles fuse with lysosomes. Lysosomes are cell organelles which contain a variety of hydrolytic enzymes. These enzymes can rapidly degrade ingested macromolecules, including DNA (24). Thus, mammalian cells possess mechanisms for protecting themselves from the potentially perturbing influences of foreign DNA.

Despite this cellular defense mechanism, exogenous foreign DNA can, under certain circumstances, become integrated in the DNA of the recipient cell. The evidence of this has come from studies of oncogenic virus transformation of mammalian, including human, cells (25). In the case of oncogenic transformation with SV40 virus, the viral DNA is apparently physically integrated into the chromosomal DNA of the recipient cell (26). It seems probable that heritable alterations of cell morphology and biochemistry are the result of the expression of one or more viral genes. Presumably, viral DNA integration takes place by base pairing of homologous regions of host cell and viral DNA followed by genetic recombination. However, the integration of oncogenic viral DNA may represent a special case since at least one viral gene product may be required for integration (27). Other, nonviral DNA molecules, unable to supply this integration function, might integrate at a much lower frequency, if at all.

In addition to integration by genetic recombination, exogenous DNA might be stabilized in the recipient mammalian cell as an independently replicating genetic unit in the cell nucleus. Although such units are known to exist in bacteria, they have not been observed in mammalian cells. However, the cytoplasmic mitochondria of mammalian cells do contain nonchromosomal, independently replicating units of DNA. The mitochondrial

DNA replication system thus offers another possible site for stabilization of exogenous DNA.

For a human genetic defect to be repaired by administering exogenous DNA, the stabilized newly introduced DNA must be correctly expressed. That is, the new gene must be correctly transcribed into mRNA and this mRNA must be correctly translated into protein. Since little is known about the regulation of mRNA synthesis and translation during natural gene expression in mammalian cells, a corresponding high degree of uncertainty exists concerning the ability of newly introduced DNA to be expressed correctly.

A variety of attempts have been made to demonstrate DNA-mediated modification of genetically mutant mammalian cells, both in vivo and in vitro. Apparently successful results in vitro have been reported for diploid human cells lacking the purine "salvage" enzyme hypoxanthine-guanine phosphoribosyltransferase (HGPRT); in human reticulocytes synthesizing an abnormal hemoglobin; in several malignant cells of mouse origin carrying markers for drug resistance; and in mouse cells with defective melanin synthesis, among others (28). In addition, transient expression of HGPRT enzyme function has been detected in human cells deficient in HGPRT after exposing them to DNA from cells with normal amounts of HGPRT (29). This suggests that exogenous DNA may be taken up and expressed, without necessarily being stabilized. However, none of the successful experiments described to date have been reproducible.

There may be several reasons for failure to demonstrate consistently the genetic modification of mammalian cells by DNA. Many previous experiments suffered from the unavailability of good genetic markers and sensitive selective systems for detecting modified cells. An important difficulty in using bulk DNA isolated from human (or other mammalian) cells is that the fraction of this DNA which is specific for any given gene is estimated to be extremely small, of the order of 10^{-7} (30). As we mentioned earlier, nonviral exogenous DNA may not be able to integrate into the chromosomal DNA of the recipient cell, thus preventing permanent genetic modification.

In spite of the lack of reproducible success in past experiments, several recent technical developments have suggested new ways in which the problems of low DNA specificity, failure of integration, and intracellular DNA degradation might be overcome.

The prospects for directing genetic modification of mammalian cells would almost certainly be enhanced by using DNA preparations containing only the gene for which the genetically defective cells are mutant. As already pointed out, both the isolation of a specific group of bacterial genes and the complete chemical synthesis of a single gene were reported recently (5, 6).

The RNA-dependent DNA polymerase recently found in RNA tumor viruses (31) could also be used for gene synthesis in vitro. Since this enzyme is able to make DNA copies from an RNA template, it offers a method for synthesizing the DNA for any specific RNA which might be isolated in pure form. Thus, it seems probable that our developing ability to isolate specific genes, or synthesize them, will eventually eliminate the problem of low specificity of the exogenous DNA.

Some workers are developing techniques which could be used to overcome the problem of stabilizing the incoming exogenous DNA in the recipient cell (32). They plan to make use of the ability of the DNA from SV40 virus to integrate into the chromosomal DNA of the cell. Specific genes will be attached to the viral DNA by means of several biochemical steps which are already known and fairly well characterized. These operations would create a hybrid DNA molecule which would carry the information for integration from the original viral DNA and perform the specific gene functions of the attached DNA. In this approach, DNA integration would be combined with biochemical manipulation of the DNA gene substance in vitro, and any gene-specific DNA segments obtained by synthesis or isolation could be utilized. It is clear that before such hybrid DNA molecules could be used in a human therapeutic situation, the oncogenic potential of the viral DNA would have to be eliminated.

In another experimental approach, virus-like particles which contain pieces of cellular DNA (pseudovirions) instead of viral DNA are being used as the vector for DNA-mediated genetic modification (33). This might help to protect the incoming DNA from intracellular degradation. However, pseudovirion DNA is probably a random collection of cellular DNA fragments (34) and hence nonspecific for any given gene; it might also be unable to stabilize itself by integration into the host cell DNA. This may explain why attempts to modify thymidine kinase–deficient mouse cells in vitro by means of polyoma virus pseudovirions have been unsuccessful (35).

It has recently proved possible to reconstruct infectious particles of several plant and bacterial viruses from the nu-

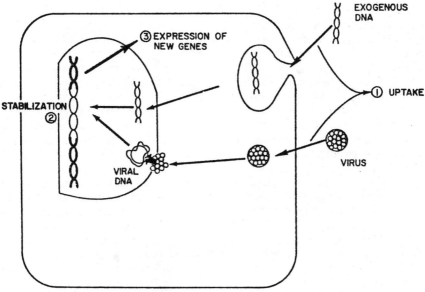

Fig. 2. Steps in the genetic modification of a mammalian cell. The added exogenous genetic information may be integrated into the chromosome of the recipient cell and become expressed as a new gene product.

cleic acid and capsid protein components (*36*). This suggests the possibility of creating artificial pseudoviruses as vectors for DNA-mediated genetic modification. These pseudovirions would contain specific DNA segments (either isolated or synthesized) surrounded by virus capsid protein. The probability of introducing a specific piece of genetic material might be greatly increased when compared with natural pseudovirions carrying randomly excised pieces of DNA. This in no way solves the difficulties of integration and expression of the genetic material. Since the specificity of virus-cell interactions is determined at least in part by the virus capsid protein, encapsidation of specific DNA molecules might confer some cell or tissue selectivity upon the DNA molecules used for gene modification.

DNA-Mediated Gene Therapy

In attempting to envision how DNA might be used as a mediator for the modification of genes in a human being suffering from a genetic defect, we foresee several kinds of new problems. First, the existence of differentiation and cell specialization in the human body will pose several questions. Many human genes are active or expressed only in a small fraction of the cells of the body. For example, the activity of the enzyme phenylalanine hydroxylase (deficient in individuals with phenylketonuria) is demonstrable only in the liver. For prospective gene therapy there might be several consequences. (i) The introduction of, for example, the gene for phenylalanine hydroxylase into cells which do not normally express this enzyme would yield no therapeutic benefits if the expression of the newly introduced genes were also blocked. Methods would have to be developed to deliver the exogenous DNA

to the appropriate "target tissue," and to confine its action solely to that tissue. (ii) Some gene products (hormones, for example) are made and secreted by one specialized group of cells and act on target cells elsewhere in the body. Synthesis and secretion of hormones such as insulin are regulated by mechanisms which are still imperfectly understood. Thus, the introduction of new genes for insulin into cells not appropriately differentiated to provide the correct synthetic and secretory responses would be of little use as a treatment for diabetes. (iii) In several genetic disorders, genetic modification of the brain cells themselves may be required to reduce the accumulation of metabolites in the brain, because the blood-brain barrier might prevent enzymes made in other parts of the body from entering the brain (*15*). We wonder whether direct genetic modification of brain cells could be made safe enough for use in human patients.

Second, regulation of the quantitative aspects of enzyme production may present a problem. By mechanisms as yet unknown, concentrations of cellular enzymes are regulated so that neither too much nor too little enzyme is produced by normal cells. How will we ensure that the correct amount of enzyme will be made from the newly introduced genes? Will the integration event, linking exogenous DNA to the DNA of the recipient cell, itself disturb other cellular regulatory circuits?

Third, the patient's immunological system must not recognize as foreign the enzyme produced under the direction of the newly introduced genes. If this occurred, the patient would form antibodies against the enzyme protein, perhaps nullifying the intended effects of the genetic intervention. This suggests that the new gene introduced during gene therapy would have to code for an enzyme with the same amino acid sequence as the human enzyme.

In addition, administration of foreign genetic material to patients carries a risk of altering the germ cells as well as the desired target cells. One might think that this problem could be circumvented by first removing some of the patients' cells, carrying out DNA-mediated genetic modification in vitro, and then reimplanting the altered cells back into the patient. However, this approach is likely to be limited by the tendency of cells to dedifferentiate and become malignant when grown in vitro.

For an acceptable genetic treatment of a human genetic defect, we would require that the gene therapy replace the functions of the defective gene segment without causing deleterious side effects either in the treated individual or in his future offspring. Years of work with tissue cultures and in experimental animals with genetic defects will be required to evaluate the potential side effects of gene therapy techniques. In our view, solutions to all these problems are needed before any attempt to use gene therapy in human patients could be considered ethically acceptable.

We are aware, however, that physicians have not always waited for a complete evaluation of new and potentially dangerous therapeutic procedures before using them on human beings. Consider how little was known of the basic aspects of virology during Jenner's development of vaccination against smallpox. In this regard, potential gene therapy techniques resemble other medical innovations. There is currently, and there may continue to be, a tendency to use incompletely understood genetic manipulative techniques, borrowed from molecular biology, in clinical settings. We believe that the first attempt at gene therapy in human patients (8) illustrates this contention.

The case in question (8) concerns two children suffering from hyperargininemia, a hereditary deficiency of the enzyme arginase. The arginase deficiency leads to high concentrations of arginine in the children's blood and cerebrospinal fluid, and has associated with it severe mental retardation (37). An attempt has been made to correct this defect at the genetic level by injecting Shope papilloma virus into the children (8). The scientific rationale for this treatment is based upon the report that the synthesis of arginase is stimulated in rabbit skin infected with Shope papilloma, and that this new arginase activity had some properties which are different from those of the normal enzyme of rabbit liver (38). In 1958, when these experiments were first reported, it was postulated that the viral DNA carried the gene for a viral arginase different from the cellular enzyme. In addition, the serums of laboratory workers who had worked with and thus been exposed to Shope papilloma virus were tested, and 35 percent of them exhibited lower concentrations of arginine than control hospital patients who had not knowingly been exposed to the virus (39). Thus, there were some grounds for believing that inadvertent infection with Shope papilloma in humans could lower the concentration of serum arginine without apparent harmful effects.

More recently, the interpretation that Shope papilloma virus codes for an arginase has been seriously questioned (40). It now appears more probable that the virus infection stimulates the production of a cellular arginase. Whether the induced arginase is coded for by viral or by cellular genes is important to the rationale of this attempt at gene therapy. If virus infection induces the synthesis of cellular arginase, and if the children have hereditarily lost the ability to produce arginase, then infecting the children with Shope papilloma virus may not have any possibility of correcting their condition (*41*).

The use of intact viruses as vectors in gene therapy raises further questions. When applied to the skin of rabbits Shope papilloma virus induces skin papillomas, a variable proportion of which develop into cancerous skin lesions. Although Shope papilloma has not had any known harmful effects on humans, tests to establish the safety of large doses have not been performed. It should also be shown that a vector for clinical gene therapy is free from other contaminating viruses latent in the cells used to produce the injected virus.

The clinical results of this therapeutic attempt are not yet known. But we are concerned that this first attempt at gene therapy, which we believe to have been premature, will serve as an impetus for other attempts in the near future. For this reason, we offer the following considerations as a starting point for what we hope will become a widespread discussion of appropriate criteria for the use of genetic manipulative techniques in humans.

Some Preliminary Criteria

We propose the following ethico-scientific criteria which any prospective techniques for gene therapy in human **patients** should satisfy:

1) There should be adequate biochemical characterization of the prospective patient's genetic disorder. It should be determined whether the patient (i) is producing a mutated, inactive form of the normal protein; (ii) is producing none of the normal protein; or (iii) is producing the normal protein in normal amounts, but the protein is rendered inactive in some way. For example, alterations in membrane structure leading to loss of the cellular receptors for insulin could produce a diabetes-like condition, even though the patient were producing normal amounts of insulin. We anticipate that defects of this type may be found affecting the activity of enzymes which are normally constituents of cell membranes. Our point is that only in the first type of genetic defect (i) would currently envisioned gene therapy techniques be likely to improve the patient's condition.

2) There should be prior experience with untreated cases of what appears to be the same genetic defect so that the natural history of the disease and the efficacy of alternative therapies can be assessed. Thus, the first reported cases of a new human genetic disease would seldom be candidates for attempts at gene therapy. The reason for this criterion comes from our accumulating experience with some of the better studied genetic defects such as phenylketonuria and galactosemia. We now observe heterogeneity in these conditions; that is, what appears to be the same genetic disease turns out to have different genetic bases in different individuals. Widespread screening for phenylketonuria in newborns has detected individuals who, like phenylketonurics, have high concentrations of phenylalanine in the serum just after birth, but have concentrations in the normal range several months later (*2*). It is now also clear that some individuals with high concentrations of phenylalanine in the serum have normal intelligence quotients (*2*). We anticipate that other genetic diseases will exhibit the same kind of heterogeneity. Concern for the welfare of each individual patient dictates that we not rush in with gene therapy until we are very sure about

the precise nature and consequences of his genetic defect.

3) There must be an adequate characterization of the quality of the exogenous DNA vector. This will require the development of new, more accurate methods of analyzing the base sequence of the DNA, if synthetic DNA molecules are to be used, or the development of new methods of isolation and purification, if naturally occurring DNA molecules are to be used. We visualize the Food and Drug Administration, or some similar organization, establishing and enforcing quality standards for DNA preparations used in gene therapy.

4) There should be extensive studies in experimental animals to evaluate the therapeutic benefits and adverse side effects of the prospective techniques. These tests should include long-term studies on the possible induction of cancer and genetic disturbances in the offspring of the treated animals. This will require the development of animal models for human genetic diseases. Previous work, which led to the isolation of a mouse strain deficient in the enzyme catalase (42) suggests that such animal models could be developed and might yield answers to some of the questions we have raised.

5) For some genetic diseases, the patient's skin fibroblasts grown in vitro reflect the disorder. Thus, in some cases it would be possible to determine whether the prospective gene therapy technique could restore enzyme function in the cells of the prospective patient. This could be done first in vitro, without any of the risks of treating the whole patient. Some side effects, such as chromosome damage and morphological changes suggesting malignancy, could also be assessed at this time. Only when a potential gene therapy technique had satisfied all these safety and efficacy criteria would it

be considered for use in human patients.

These criteria omit some other considerations which we believe are important. Although the ethical problems posed by gene therapy are similar in principle to those posed by other experimental medical treatments, we feel that the irreversible and heritable nature of gene therapy means that the tolerable margin of risk and uncertainty in such therapy is reduced. Physicians usually arrive at a judgment regarding the ethical acceptability of an experimental therapy by balancing the risks and consequences of different available treatments against their potential benefits to the patient. In general, the degree of risk tolerated in medical treatment is directly related to the seriousness of the condition.

High-risk treatments are sometimes considered more justified in life-threatening situations. For different human genetic diseases, the severity of the problem in the untreated condition and the response to currently available therapy varies greatly. Thus, phenylketonuria leads to mental retardation, but not death, in most untreated affected individuals, but the mental retardation can be avoided for the most part by prompt neonatal dietary therapy. In contrast, in the infantile form of Gaucher's disease, a deficiency in the enzyme glucocerebrosidase (important in the metabolism of brain glycolipids) leads to severe and progressive neurologic damage and death within 1 or 2 years (38). There is as yet no effective therapy. Thus, the specific characteristics of each genetic disease will be an important factor in evaluating whether or not to attempt gene therapy. We believe that the prospective use of gene therapy will need to be evaluated on a case by case basis.

Another ethical ideal which guides experimental medical treatments is informed consent. By informed consent we mean that the patient, after having the nature of the proposed treatment and its known and suspected risks explained to him by the physician, freely gives the physician his consent to proceed with the treatment. Since many of the cases where gene therapy might be indicated will involve children or newborns as patients, there will be especially troubling problems surrounding informed consent. Parents of newborn children with genetic defects may be asked to give "consent by proxy" for gene therapy. Clearly, until we know much more about the side effects of gene therapy, it will not be possible to provide them with adequate information about risks to the treated individual and his offspring.

Control of Gene Therapy

How can gene therapy in humans be controlled to avoid its misuse? By misuse we mean the premature application of techniques which are inadequately understood and the application of gene therapy for anything other than for the primary benefit of the patient with the genetic disease. In our view, it will be possible to control the procedures used for gene therapy at several levels. For example, between the patient and physician, we can usually rely upon the selection of a therapeutic technique having optimal chances of success. In general, we believe that the doctor will not recommend and the patient will not accept an uncertain, risk-laden gene therapy if a reasonably effective alternative therapy is available. However, the physician, in this as in other cases of experimental therapeutic techniques, has a near monopoly on the relevant facts about risks and benefits of various treatments. Since the physician concerned may also be active in trying to develop the gene therapy technique, how can the patient be protected from a physician who might be overeager to try out his new procedure?

It seems to us that significant opportunities for control also exist at the level of the hospital committees responsible for examining experimental techniques. Already at accredited hospitals, all proposals for research in which human subjects will be used must pass through a review committee. Further control exists through scrutiny of the proposed techniques by the physician's immediate peers.

Procedures to be used for gene therapy might also be controlled by the committees and organizations approving and funding research grants. Moderately large amounts of money will be required for the development of gene therapy techniques, hence there should be competition for public funds with other urgent medical needs. Thus, the first use of gene therapy in human patients would, of necessity, have secured the implied or direct approval of several larger public bodies beyond the principal physician-investigator. In our judgment, these levels of control will probably prove adequate to prevent misuse of projected gene therapy if, as we suspect, gene therapy is attempted in only a small number of instances. Any potential large-scale use of gene therapy (for example, the prospect of treating the approximately 4 million diabetics in the United States with DNA containing the gene for insulin) might appreciably affect the overall quality of the gene pool and would require other forms of control.

Conclusions

In our view, gene therapy may ameliorate some human genetic diseases in the future. For this reason, we believe that research directed at the development of techniques for gene therapy should continue. For the foreseeable future, however, we oppose any further attempts at gene therapy in human patients because (i) our understanding of such basic processes as gene regulation and genetic recombination in human cells is inadequate; (ii) our understanding of the details of the relation between the molecular defect and the disease state is rudimentary for essentially all genetic diseases; and (iii) we have no information on the short-range and long-term side effects of gene therapy. We therefore propose that a sustained effort be made to formulate a complete set of ethicoscientific criteria to guide the development and clinical application of gene therapy techniques. Such an endeavor could go a long way toward ensuring that gene therapy is used in humans only in those instances where it will prove beneficial, and toward preventing its misuse through premature application.

Two recent papers have provided new demonstrations of directed genetic modification of mammalian cells. Munyon et al. (44) restored the ability to synthesize the enzyme thymidine kinase to thymidine kinase–deficient mouse cells by infection with ultraviolet-irradiated herpes simplex virus. In their experiments the DNA from herpes simplex virus, which contains a gene coding for thymidine kinase, may have formed a hereditable association with the mouse cells. Merril et al. (45) reported that treatment of fibroblasts from patients with galactosemia with exogenous DNA caused increased activity of a missing enzyme, α-D-galactose-1-phosphate uridyltransferase. They also provided some evidence that the change persisted after subculturing the treated cells. If this latter report can be confirmed, the feasibility of directed genetic modification of human cells would be clearly demonstrated, considerably enhancing the technical prospects for gene therapy.

References and Notes

1. V. A. McKusick, *Mendelian Inheritance in Man* (Johns Hopkins Press, Baltimore, ed. 3, 1971).
2. D. Y. Y. Hsia, *Progr. Med. Genet.* 7, 29 (1970).
3. E. R. Kramm, M. M. Crane, M. G. Sirken, M. D. Brown, *Amer. J. Public Health* 52, 2041 (1962).
4. V. A. McKusick, *Annu. Rev. Genet.* 4, 1 (1970).
5. J. Shapiro, L. MacHattie, L. Eron, G. Ihler, K. Ippen, J. Beckwith, *Nature* 224, 768 (1969).
6. K. L. Agarwal, H. Buchi, M. H. Caruthers, N. Gupta, H. G. Khorana, K. Kleppe, A. Kumar, E. Ohtsuka, U. L. Rajbhandary, J. H. Van De Sande, V. Sgaramella, H. Weber, T. Yamada, *ibid.* 227, 27 (1970).
7. S. Rogers, *New Sci.* (29 Jan. 1970), p. 194; H. V. Aposhian, *Perspect. Biol. Med.* 14, 98 (1970).
8. *New York Times*, 20 Sept. 1970.
9. A. G. Schwartz, P. R. Cook, H. Harris, *Nature* 230, 5 (1971).
10. B. D. Davis, *Science* 170, 1279 (1970).
11. J. B. Stanbury, J. B. Wyngaarden D. S. Fredrickson, Eds., *The Metabolic Basis of Inherited Disease* (McGraw-Hill, New York, ed. 2, 1966).
12. N. A. Holtzman, *Annu. Rev. Med.* 21, 335 (1970).
13. W. N. Kelley, F. M. Rosenbloom, J. Miller, J. E. Seegmiller, *N. Engl. J. Med.* 278, 287 (1968).
14. C. A. Mapes, R. L. Anderson, C. C. Sweeley, R. J. Desnick, W. Krivit, *Science* 169, 987 (1970).
15. H. L. Greene, G. Hug, W. K. Schubert, *Arch. Neurol.* 20, 147 (1969).
16. G. Hug and W. K. Schubert, *J. Cell Biol.* 35, C1 (1967).
17. A. Milunsky, J. W. Littlefield, J. N. Kanfer, E. H. Kolodney, V. E. Shih, L. Atkins, *N. Engl. J. Med.* 283, 1370, 1441, 1498 (1970).
18. R. O. Brady, *Annu. Rev. Med.* 21, 317 (1970).
19. S. Pell and C. A. D'Alonzo, *J. Amer. Med. Ass.* 214, 1833 (1970).
20. R. D. Hotchkiss and M. Gabor, *Annu. Rev. Genet.* 4, 193 (1970).
21. A. S. Fox, S. B. Yoon, W. M. Gelbart, *Proc. Nat. Acad. Sci. U.S.* 68, 342 (1971).
22. S. A. Aaronson and G. J. Todaro, *Science* 166, 390 (1969).
23. H. J. P. Ryser, *ibid.* 159, 390 (1968).
24. J. T. Dingle and H. B. Fell, *Frontiers Biol.* 14A, 220 (1969).
25. R. Dulbecco, *Science* 166, 962 (1969); M. Green, *Annu. Rev. Biochem.* 39, 701 (1970).
26. J. Sambrook, H. Westphal, P. R. Srinivasan,

77

R. Dulbecco, *Proc. Nat. Acad. Sci. U.S.* **60**, 1288 (1968).

27. M. Fried, *Virology* 40, 605 (1970).

28. E. H. Szybalska and W. Szybalski, *Proc. Nat. Acad. Sci. U.S.* 48, 2026 (1962); M. Fox, B. W. Fox, S. R. Ayad, *Nature* 222, 1086 (1969); R. A. Roosa and E. Bailey, *J. Cell. Physiol.* 75, 137 (1970); L. M. Kraus, *Nature* 192, 1055 (1961); D. Roth, M. Manjon, M. London, *Exp. Cell Res.* 53, 101 (1968); J. L. Glick and C. Sahler, *Cancer Res.* 27, 2342 (1967); E. Ottolenghi-Nightingale, *Proc. Nat. Acad. Sci. U.S.* 64, 184 (1969).

29. T. Friedmann, J. H. Subak-Sharpe, W. Fujimoto, J. E. Seegmiller, paper presented at Society of Human Genetics Meeting, San Francisco, Oct. 1969.

30. Calculated with the assumption that (i) the DNA contents of a diploid human cell is 4×10^{12} daltons; (ii) representative gene codes for a protein containing 200 amino acids equivalent to 4×10^5 daltons of DNA; and (iii) there are two copies of each gene per cell. This would represent a minimum estimate if the redundant DNA segments in human cells include genes which specify enzymes.

31. H. M. Temin and S. Mizutani, *Nature* 226, 1211 (1970); D. Baltimore, *ibid.*, p. 1209.

32. P. Berg, and D. M. Jackson, personal communication.

33. J. V. Osterman, A. Waddell, H. V. Aposhian, *Proc. Nat. Acad. Sci. U.S.* 67, 37 (1970).

34. L. Grady, D. Axelrod, D. Trilling, *ibid.*, p. 1886.

35. B. Hirt, seminar at Brandeis University, Oct. 1970.

36. H. Fraenkel-Conrat, *Annu. Rev. Microbiol.* 24, 463 (1970).

37. H. G. Terheggen, A. Schwenk, A. Lowenthal, M. Van Sande, J. P. Colombo, *Lancet* II-1969, 748 (1969); Z. *Kinderheilk* 107, 298 and 313 (1970).

38. S. Rogers, *Nature* 183, 1815 (1959).

39. ——— and M. Moore, *J. Exp. Med.* 117, 521 (1963).

40. P. S. Satoh, T. O. Yoshida, Y. Ito, *Virology* 33, 354 (1967); G. Orth, F. Vielle, J. B. Changeux, *ibid.* 31, 729 (1967).

41. This objection would be straightforward if only one type of arginase were present in human cells. Even though there seem to be two arginase isozymes in both human liver and erythrocytes, the objection is still cogent since one isozyme contributes 90 to 95 percent of the total arginase activity, and the two isozymes cross-react immunologically. See J. Cabello, V. Prajoux, M. Plaza, *Biochim. Biophys. Acta* 105, 583 (1965).

42. R. N. Feinstein, M. E. Seaholm, J. B. Howard, W. L. Russel, *Proc. Nat. Acad. Sci. U.S.* 52, 661 (1964).

43. H. Harris, *Frontiers Biol.* 19, 167 (1970).

44. W. Munyon, E. Kraiselburd, D. Davis, J. Mann, *J. Virol.* 7, 813 (1971).

45. C. R. Merril, M. R. Geier, J. C. Petricciani, *Nature* 233, 398 (1971).

Molecular Biology:
Gene Insertion into Mammalian Cells

By DANIEL RABOVSKY

The problem of inserting specific genes into human cells has intrigued molecular geneticists, and the prospect of the successful solution of this problem has concerned everyone. Both the excitement and the concern have grown· now that the armchair speculations—and exploratory results—of a few years ago have matured into hard experimental work. The current results of that work indicate that animal viruses, bacterial viruses, and cell fusion techniques are all capable of introducing new functional genes into mammalian cells, although many of the fundamental genetic and regulatory processes in mammalian cells remain unknown.

Much has been learned about the genetic code and the mechanisms of the replication of DNA, the transcription of DNA into RNA, and the translation of RNA into protein, especially in bacterial cells. A clever and sufficiently industrious molecular geneticist can often produce a specific mutation in any of a large number of genes in the bacterium *Escherichia coli*, can delete genes or add new ones from outside the cell, and can then regulate the expression of genetic traits inside the cell. But the extension of these techniques from bacteria and bacterial viruses (bacteriophages) to nucleated (eukaryotic) cells, especially human cells, awaited new tools and more knowledge.

Several biologists have studied the interaction of foreign DNA with nucleated cells. Among these, Pradman Qasba and Vasken Aposhian at the University of Maryland School of Medicine in Baltimore, have recently shown that one type of animal virus can be used to transport DNA from mouse cells into the nuclei of human cells. At the Roswell Park Memorial Institute in Buffalo, W. Munyon and his coworkers have shown that another type of animal virus may have inserted a specific gene into mouse cells without harming the cells. These workers found that the enzyme specified by this gene was made by the cell and that the new gene seemed to be replicated as the cells divided.

Munyon and his group infected mutant L cells (a line of mouse tissue culture cells) that lacked the enzyme thymidine kinase with the animal virus

SCIENCE, 1971, Vol. 174, pp. 933-934. Copyright 1971 by the American Association for the Advancement of Science.

herpes simplex. The virus had been irradiated with ultraviolet light to decrease its ability to kill cells (*1*). Herpes simplex virus normally induces a thymidine kinase activity during infection before it kills the cells, but in this experiment about 0.1 percent of the infected L cells were transformed by the irradiated virus into stable cells that had thymidine kinase activity and were maintained in culture for 8 months. No measurable proportion ($< 10^{-8}$) of control L cells gained the ability to express thymidine kinase when uninfected cells or cells infected with a herpes simplex mutant that does not induce thymidine kinase activity were examined.

These results are consistent with the idea that the herpes simplex virus introduced a gene for thymidine kinase into the L cells and that this gene was then maintained and replicated by the cells. However, Munyon notes the possibility that a herpes gene product may have simply induced the stable expression of a gene that was already present in the L cells.

Aposhian has proposed that pseudovirions—which consist of normal virus' protein coats that have enclosed foreign pieces of DNA—might be able to deliver foreign genes into another cell, and that these new genes could function and be replicated in the cell. Qasba and Aposhian have now established that pseudovirions of an animal virus, polyoma, containing labeled DNA from mouse embryo cells can deliver this DNA inside the nuclei of human embryo cells in the form of uncoated pseudovirion DNA as soon as 24 hours after infection of the human cells by the polyoma pseudovirions (*2*). At present no actual expression or replication of any newly introduced genes has been shown in this system, but many workers think it likely that uncoated mammalian DNA in a mammalian cell nucleus is capable of integrating itself into the host chromosome and functioning in some way.

Carl Merril and his co-workers (*3*) at the National Institutes of Health in Bethesda have taken a different approach. They have shown that a bacteriophage is capable of introducing a selected functional gene into human cells. Merril had worked with the bacterial virus lambda phage, one of the transducing phages of *E. coli*. Transducing phages have been among the most useful genetic tools available to molecular biologists. Sometimes bacterial genes are included in the DNA of the new phage produced during infection of a bacterium by a transducing phage. If these phages infect another bacterium they can transduce the bacterial genes that they carry into their new host. In this way transducing phages are often used to selectively introduce new functional genes into bacteria.

The genetic structure of lambda has been very well mapped, and a number of remarkable lambda transducing phages (that is, lambda that carry bacterial genes) are available. Merril's group decided to attempt the transduction of cultured human fibroblast cells from the skin of a patient with galactosemia, the disease that results from an inborn error of metabolism in which the enzyme α-D-galactose-1-phosphate uridyl (GPU) transferase is lacking. The cultured fibroblasts were treated with lambda phage (λ pgal) that carried the *E. coli* galactose operon—a set of genes that codes for the enzymes which convert galactose to glucose and controls their synthesis.

Merril and his co-workers hoped that the GPU transferase gene, which is part of the galactose operon, could be supplied to the fibroblasts by λ pgal and that the fibroblasts would use this transduced gene to make the GPU transferase enzyme.

These expectations were fulfilled. Infection of the fibroblasts by λ pgal resulted in both the production of lambda-specific RNA and in the appearance of GPU transferase activity in the fibroblasts. The lambda-specific RNA and the new GPU transferase activity were found in significant amounts and persisted at the same amounts per cell for more than 40 days (during which more than eight doublings of the cells took place). In addition, uncoated λ pgal DNA was, in Merril's experiment, at least as effective as the whole virus particle. Control infections by normal lambda and by λ pgal with a mutation that inactivates its transferase resulted in the production of lambda-specific RNA but not in transferase activity.

There is still no evidence to indicate what part of the cell houses the lambda DNA. But the experiments performed by Merril's group have shown that lambda DNA can enter human cells, and once there at least some of its genes can be replicated, transcribed, and translated.

Cell Fusion Techniques

Viruses are not the only means of introducing new genes into mammalian cells, however. Henry Harris' group at Oxford in England has now shown that cell fusion can also be used to insert a functional gene from chicken cells into mouse tissue culture cells (4). If two cells are fused or if another nucleus is introduced into a cell, the dividing time of the resulting multinucleate cell is primarily determined by the nucleus that is closest to division. The other nuclei may be forced to divide before they are ready, and their chromosomes usually undergo a "premature condensation" which results in their fragmentation or "pulverization."

The Harris group fused chick red blood cells, whose nuclei carry the gene for chick inosinic acid pyrophosphorylase, with mouse fibroblast A_9 cells, which are a mutant line of mouse L cells deficient in this enzyme. During cell division, the nuclear membranes in these hybrid cells disappear, and the chick chromosomes undergo pulverization. Although only mouse nuclei appeared in the daughter cells after mitosis, some (2×10^{-5}) of the daughter cells had gained the ability to synthesize chick inosinic acid pyrophosphorylase. However, after more than 100 generations in culture, 20 percent of these transformed hybrid cells lost the ability to make this enzyme, as compared to 1 out of 10^6 in a normal L cell population. Hence the chick gene for inosinic acid pyrophosphorylase was replicated and remained functional in the mouse cells, although it was not as genetically stable as the normal mouse gene for this enzyme. This experiment did not provide direct evidence for the location of the chick gene inside the mouse cells, but indirect evidence leads the Harris group to name the mouse nucleus as its probable residence.

A number of laboratories are already extending these gene transfer techniques. Aposhian's group is now studying the genetic properties of polyoma pseudovirions in mice. They are also attempting to produce polyoma pseudovirions containing the gene for human thymidine kinase. Merril has infected whole animals with transducing lambda bacteriophage in order to determine whether any new traits gained by the animals from the genes that are transported by the phage can be inherited from one generation to the next.

The ability to transfer small segments of DNA from one cell to another gives molecular geneticists the opportunity to map the locations of genes within mammalian chromosomes. In turn, understanding the method by which genes are organized into functional units within the chromosome may reveal how the regulation of gene expression takes place. This regulation may play an important role in the differentiation of animal cells.

Current developments certainly do not yet add up to "genetic engineering"; but there now exists very strong evidence, with a number of different techniques, that experimenters can transfer genes between, and insert them into, mammalian cells—a development that opens to experiment many questions about gene function in animals and in humans. Now that a few of the basic tools of molecular genetics have been extended to mammalian cells, more than a few biologists share Aposhian's concern that science "will give us gene therapy before society is prepared for it."

References

1. W. Munyon, E. Kraiselbrud, D. Davis, J. Mann, *J. Virol.* **7**, 813 (1971).
2. P. K. Qasba and H. V. Aposhian, *Proc. Nat. Acad. Sci. U.S.* **68**, 2345 (1971).
3. C. R. Merril, M. R. Geier, J. C. Petricciani, *Nature* **233**, 398 (1971)
4. A. G. Schwartz, P. R. Cook, H. Harris, *Nature New Biol.* **230**, 5 (1971).

REPRODUCTIVE ENGINEERING: INTERVENTION IN DEVELOPMENT

GENETIC ENGINEERING PORTENDS A GRAVE NEW WORLD

Caryl Rivers

A society matron proudly introduces her two sons and her daughter. They are her children by every biological rule but one: She has never been pregnant. Sex cells from her body and her husband's body were implanted three times in the womb of a "proxy mother," who was paid a union wage to carry each fetus and give birth.

An astronaut is carried aboard a space vehicle destined to probe the outer limits of this galaxy. He has no legs. Legs would be only an inconvenience during the years in which the astronaut will be confined to his spaceship. For this reason he was programed to be born legless.

A Latin American dictator has some skin tissue scraped from his left arm. Nine months later, five hundred babies emerge from a factory that contains five hundred artificial wombs. The babies are genetic carbon copies of the dictator.

Although fiction today, these events could become realities with relatively slight advances in man's most adventurous and morally complex science, genetic engineering.

This new science is experimenting with a technique that would make possible the manipulation of an embryo during gestation so as to change its physical characteristics. It may offer an alternative to natural human reproduction—a process that would allow the implantation of a fertilized ovum in the womb of a "host" mother or in an artificial womb inside a laboratory. Not too far off, according to specialists in the field, is the possibility of creating children with only one parent who will be biological duplicates of that single parent. Genetic engineering, in short, is on the brink of revolutionizing the traditional concepts of man, God, and creation.

The moral questions posed by recent advances in this and related life sciences are no longer speculative. They have to be faced as practical realities, and, in parts of the scientific community, this is now being done. Indeed, a heated debate is in progress. Some scientists encourage genetic experimentation, putting their faith in man's rational power. They say, in effect, that the only thing too sacred to tamper with is scientific investigation itself. Others, however, point to numerous instances in which technology has outrun man's wisdom (nuclear stockpiles, for example), and they warn that if guidelines in genetic technology are not immediately set forth, geneticists may bring into being a terrifying, uncontrollable Huxleian nightmare.

SATURDAY REVIEW, April 8, 1972, pp. 23-27.

83

Those who urge restraint say it is past time to start thinking about controls for genetic technology. Scientific advances have already made it possible to change fundamentally the human reproductive process. The Caesarean section is now a common operation and constitutes a basic tampering with the way in which babies are born. There have already been successful blood transfusions to unborn children in cases of Rh blood incompatibility. An estimated 25,000 women whose husbands are sterile resort to artificial insemination in this country each year, and more than a third of them give birth as a result.

Test-tube fertilization of a human ovum is already a reality. In 1950 Dr. Landrum Shettles of Columbia University said he had achieved in vitro —outside the womb—fertilization of an ovum and had maintained the embryo's life for six days. In 1961 Dr. Daniele Petrucci of the University of Bologna claimed he had fertilized an egg in vitro and that the embryo lived for twenty-nine days, although it became enlarged and deformed. His work was condemned by Pope John XXIII.

Perhaps the most advanced work in the field is being done at Cambridge University in England by Drs. Robert G. Edwards and P. S. Steptoe. They have produced the best evidence to date of true in vitro fertilization and are presently attempting to implant the fertilized egg in the uterus of the donor. They obtain the ova for their experiments from volunteer women who are seeking to overcome infertility. One woman, for example, had blocked Fallopian tubes that made normal conception impossible.

Once an ovum can be successfully implanted in the womb, the way is open for "proxy mothers." A fertilized ovum could be removed from the womb of a woman after she had conceived and then be implanted in the uterus of another woman who would ultimately give birth. The child, of course, would carry the genetic identity of its true parents. The proxy mother would be only a temporary host, with no genetic relationship to the child. Technology could go one step further and simply eliminate the process of pregnancy and childbirth altogether. A human embryo could be removed from the uterus and placed in an artificial womb. A prototype of such a device was constructed by Dr. Robert Goodlin at Stanford University. It is a pressurized steel chamber containing an oxygen-rich saline solution that would push the oxygen through the body of the fetus. A fetus put into the chamber, however, would perish from its own wastes, which the placenta draws off in natural pregnancy. Dr. Goodlin is doing no further work in the area and has announced that he does not intend to resume his experiments with the artificial womb.

At the National Heart Institute at Bethesda, Maryland, Drs. Warren Zapol and Theodore Kolobow are placing lamb fetuses in a liquid solution and attaching the umbilical cords to machinery that contains an artificial lung, a pump, and a supply of nutrients. The fetuses survive several days before "dying" of cardiac arrest.

Such ex utero gestation might offer some obvious advantages. If made common practice, doctors would be able to monitor the growth of the embryo closely. A new specialist, the "fetologist," would check to ensure normal development and catch possible medical problems in the earliest stages. The trauma of birth, from which some babies die or suffer injury, would be avoided altogether. Ex utero development would make more likely manipulation of the developing fetus to change its sex, size, and other physical characteristics.

French biologist Jean Rostand speculated that, once the size of the human brain is not limited by the size of the female pelvis, it might be possible to double the number of fetal brain cells. Would this produce superintelligent people or monstrous misfits? Where will the line be drawn between legitimate medical practices and a Frankenstein-like toying with the human condition? And who will draw the line?

Another possibility of genetic engineering that raises serious moral questions is the creation of what are called clones—carbon copies of a human being already in existence, resulting from a process much more revolutionary than the merging of sperm and egg in a test tube.

Normally, the sexual process by which life begins guarantees the diversity of the species. The child inherits characteristics from both the mother and the father so that, while he may resemble either or both of them, he is unique. There are two kinds of cells in the human body—sex cells (ova and sperms) and body cells. Each body cell has a nucleus containing forty-six characteristic-determining chromosomes. Sex cells have only half that number. A merger of two sex cells is required for the resulting zygote to have a complete set of chromosomes and to begin dividing.

But human reproduction could take another route. Body cells contain the requisite number of chromosomes to start a new individual, but they cannot divide the way the fertilized egg does. The body cells have very specialized jobs to do. Some go into making hair, others teeth, others skin, and so forth. Only the part of the body-cell mechanism that is useful in its specific job gets "switched on." If a body cell could receive a signal to switch on all of its mechanisms, it would divide and subdivide and multiply, finally developing into a new human being.

Accordingly, if such a "switching on" technique were developed, a body cell could be taken from a donor—scraped from his arm, perhaps—and be chemically induced to start dividing. The cell could be implanted in an artificial womb or in the uterus of a woman, where, presumably, it would develop like a normal fetus. The baby would be genetically identical to the donor of the cell—his twin, a generation removed. It would have only one true parent. Its "mother" would be, like the proxy mother, only a temporary host.

The possibilities of human cloning are enough to startle even a science-fiction writer. Societies would be tempted to clone their best scientists, soldiers, and statesmen. An Einstein might become immortal through his carbon copies. Or a Hitler. What better way for a dictator to extend his power beyond the grave? A carbon copy might be the ultimate expression of human egoism.

Clones have already been produced from carrots and frogs. Dr. J. B. Gurdon of Oxford University produced a clone from an African clawed frog by taking an unfertilized egg cell from a frog and destroying its nucleus by radiation. He then replaced it with the nucleus of a cell from the intestines of a tadpole. The egg began to divide as if it had been normally fertilized. The result: a twin of the donor tadpole.

Some scientists estimate that human cloning will be possible by the end of the 1970s. If human cloning were ever practiced on a wide scale, it would drastically affect the course of human evolution. Nobel Prize-winning geneticist Dr. Joshua Lederberg has said he can imagine a time in the future when "clonishness" might replace the present dominant patterns of nationalism and racism.

Cloning is not the only avenue man might take to redesign himself. Advances in molecular biology could lead to genetic surgery: the addition of genes to or removal of them from human cells. In experiments with mouse cells, scientists added healthy new genes to a defective cell. The new genes not only corrected an enzyme deficiency in the cell but were duplicated when healthy new cells began to divide. It is now possible, in principle, to remove defective cells from a human patient, to introduce new genes, and to place the "cured" cells back into the patient.

Genetic surgery might also be put to more ambitious uses. J. B. Haldane, the late, renowned British geneticist, was one of the first to speculate and

85

predict that men would tailor-make new kinds of people for space travel. He suggested the legless man referred to earlier, who would be ideally suited to living in the cramped quarters of a space capsule for long journeys; men with prehensile feet and tails for life on asteroids where low gravity makes balance difficult; and muscular dwarfs to function in the strong gravitational pull of Jupiter.

Genetic engineering might also bring to life one of the mixtures of man and animal known in all the myths of man: the Minotaur, the Centaur, and the Gorgons. So far, however, human and animal chromosomes have been successfully mingled only in tissue-culture experiments.

Modern medicine may be able to produce not only man-animal combinations but a man-machine combination for which a name has already been coined: cyborg. We are familiar with the use of artificial material in the human body: plastic arteries, synthetics to replace damaged bone, or even an artificial heart. Who will draw the dividing line between man and robot?

There are few existing guidelines for the manipulation of "living" material and human beings. Reports from Britain describe experiments in which pregnant women awaiting abortion have been exposed to certain kinds of sound waves to determine if there will be extensive chromosome damage in the fetus. Is it immoral to damage a fetus for experimental purposes, even though it will be aborted? Does an embryo "grown" in a laboratory have any legal rights?

Even the practice of genetic counseling—guidance based upon an individual's chromosomal make-up—which is becoming increasingly common, raises ethical questions. Would an individual who is marked as a carrier of deleterious genes carry a social stigma as well? Dr. Marc Lappé of the Institute of Society, Ethics, and the Life Sciences, a group set up to examine the moral questions posed by

science, thinks he would.

"It could lead to a subtle shift in the way we identify people as human," he says. "We could say, 'Oh, he has an extra chromosome.' We could then identify him as being qualitatively different. He is somewhat imperfect, perhaps not as human as the rest of us."

Suppose, for example, a genetic counselor knows that a newborn baby has an extra Y chromosome. There is evidence that the extra Y chromosome predisposes an individual to aggressive behavior. Should that become part of his medical record? What impact might that knowledge have on a school principal if a child with an extra Y becomes involved in typical childish pranks? Should the parents know, or would the knowledge have an adverse effect on the way they bring up the child?

The agonizing moral question posed by almost all aspects of genetic engineering has divided the scientific community. Hearing the arguments among scientists, one is reminded of the story about two men in a jail cell. One looked out and saw the mud; the other saw the stars.

Dr. Robert Sinsheimer of the California Institute of Technology sees the promise: "For the first time in all time a living creature understands its origin and can undertake to design its future. Even in the ancient myths man was constrained by his essence. He could not rise above his nature to chart his destiny."

Dr. Sinsheimer goes on to say that those who oppose genetic engineering "aren't among the losers in the chromosomal lottery that so firmly channels our human destiny." Repugnance to advances in genetic technology "isn't the response of four million Americans born with diabetes, or the two hundred fifty thousand children born in the United States every year with genetic diseases, or the fifty million Americans whose I.Q.s are below ninety."

Dr. Salvador Luria of MIT, a Nobel laureate, takes the opposite viewpoint. "We must not ignore the possibility that genetic means of controlling human heredity will become a massive means of human degradation. Huxley's nightmarish society might be achieved by genetic surgery rather than by conditioning and in a more terrifying way, since the process would be hereditary and irreversible."

The work of Steptoe and Edwards in Cambridge has been extremely controversial among scientists. Is implantation of a fertilized ovum back into the uterus a desirable medical advance or the opening of a Pandora's box? Dr. Joseph Fletcher of the University of Virginia Medical School argued the former in a recent article that he published in the *New England Journal of Medicine*.

"It seems to me," he wrote, "that laboratory reproduction is radically human compared to conception by ordinary heterosexual intercourse. It is willed, chosen, purposed, and controlled, and surely these are among the traits that distinguish Homo sapiens from others in the animal [world], from the primates down. Coital reproduction is therefore less human than laboratory reproduction; more fun, to be sure, but with our separation of baby-making from love-making, both become more human, because they are matters of choice, not chance. I cannot see how humanity or morality [is] served by genetic roulette."

At the opposite pole, Dr. Leon Kass, the executive director of a committee on the life sciences and social policy for the National Academy of Sciences, argues that the further reproduction is pushed into the laboratory the more it becomes sheer manufacture.

"One can purchase quality control of the product only by the depersonalization of the process," he says, going on to ask, "Is there not some wisdom in the mystery of nature that joins the pleasure of sex, the communication of love, and the desire for children in the very activity by which we continue the chain of human existence? Is not human procreation, if properly understood and practiced, itself a humanizing experience?"

Dr. Kass points to the lonely depersonalized experiences old age and dying have become as a result of medical technology. The aged are kept alive but barely able to function, and they die in institutions surrounded by clacking machinery and uncaring strangers. He also believes that laboratory reproduction will deal a near fatal blow to the human family.

"The family is rapidly becoming the only institution in an increasingly impersonal world where each person is loved not for what he does, or makes, but simply because he is. Destruction of the family unit would throw us, even more than we are now, on the mercy of an impersonal, lonely present."

There are some members of the scientific community who take what might be called an "amber light" approach to genetic technology: Proceed with extreme caution.

Isaac Asimov, the biochemist who is also well known as a science-fiction writer, says we should be very sure what we're about before we start tinkering with genes. "We should intervene if we have reasonable suspicion that we can do so wisely. If our choice is doing nothing or doing something without knowing, we should do nothing. But wise change is better than no change at all. It's the same as changing the environment. Every time we build a dam there is a gain and there is a loss. What has usually happened is that we have considered the short-term gain without thinking what the long-term effect will be."

Dr. George Wald, a Nobel laureate at Harvard who has spoken out on a wide range of social issues, emphasizes the fact that every organism alive today represents an unbroken chain of life that stretches back some three billion years. That knowledge, he says, calls for some restraint. The danger he senses in genetic technology is a movement toward reducing man's unpredictability. "With animals we have abandoned natural selection for the technological process of artificial selection. We breed animals for what we want them to be: the pigs to be fat, the cows to give lots of milk, work horses to be heavy and strong, and all of them to be stupid. This is the process by which we have made all of our domestic animals. Applied to men, it could yield domesticated men."

Wald sees the trends of modern life working to erode one of the glories of being human: free will. "Free will is often inefficient, often inconvenient, and always undependable. That is the character of freedom. We value it in men. We disparage it in machines and domestic animals. Our technology has given us dependable machines and livestock. We shall now have to choose whether to turn it to giving us more reliable, efficient, and convenient men, at the cost of our freedom. We had better decide now, for we are already not as free as we once were, and we can lose piecemeal from within what we would be quick to defend in a frank attack from without."

Dr. Harold P. Green, a law professor at George Washington University, argues for the creation of an agency within the government whose sole task would be to make the case against technological advances. With any technology, he says, the benefits are immediate and obvious, and there are powerful vested interests to articulate them. The risks are more remote, and there are few people to argue these points. A prestigious new agency so mandated might right the present imbalance.

Senator Walter Mondale has proposed a fifteen-man presidential advisory committee that would spend two years looking into the moral problems posed by the life sciences. The commission would devise a structure to give society controls over such things as genetic technology. The proposal was approved by the Senate and, at last report, is languishing in a House committee.

Scientists themselves have shown growing awareness of the moral implications of their work. "Scientists have always had a supreme obligation to be concerned about the uses of their work," Isaac Asimov says. "In the past an ivory-tower scientist was just stupid. Today he's stupid and criminal."

But is it enough that scientists develop sensitivity to the moral complexities of the new technology? Can they alone grapple with choices so fundamental they could affect the future of human heredity? James Watson has called the idea that science must always move bravely forward "a form of laissez-faire nonsense dismally reminiscent of the creed that American business, if left to itself, would solve everybody's problems."

Seeing a child lying crippled by a genetic disease, one can't help thinking that, if such a scourge could be lifted from the children of the future, it would be worth the risk of any Brave New Worlds. It is probably unwise and perhaps impossible to barricade any street of scientific inquiry and say: No Admittance. But neither can we blind ourselves to the consequence of traveling that street until it is too late. Dr. Kass has summarized the question posed by genetic engineering: "Human heredity is intricate and mysterious. We must face the prospect of intervention with awe, humility, and caution. We may not know what the devil we are doing."

88

R. A. Beatty

The future of reproduction

THROUGHOUT HISTORY the mechanics of human reproduction has been rather simple. The union of man and woman has, sooner or later, produced children. No union, no children. Round this simple physiological basis we have woven our complex and varied religious, social and economic attitudes to human procreation and to sex in general. Quite suddenly, however, the material basis has begun to change. New possibilities have arisen in the technology of reproduction, possibilities that did not exist when our cultural attitudes evolved. Are we going to accommodate the new possibilities within the confines of the older frame of reference—or are we going to alter the framework?

The purpose of this article is not to forecast the future, but to review some of the discoveries in reproductive technology in relation to human beings and to consider what people think about them, for what they think will determine the future. I would like to say here and now that I do not intend to spend time on the familiar aspects of the mechanics of contraception. At last we have a suppressor of ovulation (the 'pill') which causes pregnancy to be averted before the embryo comes into being. At present it has some side-effects which, although real and important enough for some individual women, are unimportant on a population basis. A corresponding 'male pill' is being investigated; so far its side-effects still include loss of libido and even toxicity. While there is no application yet in sight for immunological

methods of neutralizing either the male or female gamete, surgical methods of contraception are of course available and voluntary sterilization of men is practised on a large scale (sterilization of women remains a major operation). And finally there is termination of pregnancy, a procedure which is distasteful in some countries but normal in others, and perfectly safe when done professionally. This said, I would now like to get on with the real subject of this article: the future of reproduction research.

IN ANIMALS the onset of 'heat'—the period of desire—and of ovulation—the shedding of unfertilized eggs from the ovary—are interconnected events, but the·nature of the connection varies with the species. Most animals have the level of their reproductive hormones determined by the season of the year. At a certain season, the hormone concentration brings about both heat and ovulation, so that the availability of fertilizable eggs, and receptivity to the male, coincide in time. But in the wild rabbit, for instance, something quite different happens. There is a long seasonally determined heat, during which eggs become ready in the ovary, but actual shedding has to be triggered off by the act of coitus. The same happens in the domestic rabbit, except that it is on heat virtually all the year.

Considerable control over the promotion of ovulation can be exerted in mammals. A simple hormone injection into a female of the laboratory rabbit

SCIENCE JOURNAL, 1970, Vol. 6, pp. 92-99.

brings about ovulation 10 hours later; here the treatment replaces the coital stimulus. If fertilized—even by artificial insemination with no coital stimulus —such eggs produce young at the normal time. All the individuals of a flock of sheep can be brought into heat at the same time by appropriate hormone treatment, so that the farmer can arrange in advance convenient dates for mating, lambing and the disposal of the offspring. And the offspring will be a uniform product of much the same age. Hormone treatment can also induce 'super-ovulation' in many animals. In this case the ovary is made to shed more than the normal number of eggs—a useful laboratory procedure for producing large numbers of eggs from as few animals as possible. But the technique is of limited use in farm animals, because there is a ceiling to the number of foetuses or young that a mother can nourish properly *in utero* or by suckling.

In human beings, the biology of ovulation is characteristically different. Traces of a heat cycle exist, but the human female is receptive to coitus at all times of year. Because it is normal for an egg or eggs to be shed once a month, there is little call for a technique that will induce ovulation timed to the week or the day. The main application in humans is a treatment of certain cases of infertility in which the whole reproductive system in the woman is normal except for one thing: a hormone deficiency that prevents her from ovulating her own eggs. This can be righted by supplying the hormone artificially, and the ovulated eggs can then be fertilized and produce normal children. Unfortunately, there is no proper control yet over the number of eggs shed; in other words, treatment may also induce superovulation. We are familiar with recent instances of fertility treatments that have led to multiple births of some half a dozen tiny babies which require a major and not always successful medical effort to save their lives. One can feel fairly confident that the treatment can soon be made less erratic. From the point of view of the population, prevention rather than promotion of ovulation is more important.

ACCESS TO EGGS and spermatozoa is essential for scientific and medical study and for the development of reproductive technology. The male presents no problem: stimulation of the male organ is usually sufficient to get an adequate supply of spermatozoa. In the female, eggs can be readily obtained from the laboratory animals by pricking the liquid filled Graafian follicle on the surface of the ovary or from the upper part of the female tract (Fallopian tube) or from the uterus (see diagram on page 54). Many years ago it was discovered that the mere act of recovering eggs from the Graafian follicle has in fact a useful side effect: it causes the immature follicular eggs to complete the first part of their maturation processes and become ready for fertilization. This automatic stimulation of the maturation process occurs in many mammals, including woman.

Human eggs, however, are much more difficult to get hold of. An operation is required and there is no question of operating on a normal woman merely to obtain eggs for research. Advantage is of course taken of occasions when an operation has to be carried out for other reasons—say, the removal of a diseased ovary or tract—and where the fate of the organs removed is not relevant to the immediate needs of the patient. In such material, Dr R. G. Edwards at Cambridge University has already solved one technical problem and for the first time unfertilized eggs have been obtained in large numbers from the ovaries of women. With the consent of the patient, it is also possible to prick the Graafian follicles or to flush out the genital tract without removing the organs.

But from all these various techniques, very few fertilized human eggs have so far been recovered. This means that no one has a proper picture of the very early stages of human development. The biological requirements of these early stages cannot receive serious medical consideration because no one knows what the requirements are. This is the main reason why doctors and scientists alike want to achieve 'test tube' development of embryos so that the requirements can be worked out. There is much need for a non-operative method of recovering human eggs. Animal eggs can be recovered in this way, but the success rate is low.

The long term storage of human spermatozoa at low temperature is already with us. Children have been born after artificial insemination (AI) of spermatozoa maintained at low temperature for many months. However, not all semen samples store satisfactorily—a technical problem that is under current investigation. Nevertheless, in man, emission of spermatozoa and fertilization of the egg are at least events which are now separable in time.

The AI experiments in human beings have been carried out with confidence because of the massive experience derived from early pilot experiments with animals which were performed on a really gigantic scale. For instance, in many parts of the world it is now routine practice to store the semen of bulls at low temperature. Cattle semen stored in this way has even maintained its fertility for as long as 12 years, and as far as is known storage could be for an indefinite time. There are no detectable differences between calves born after artificial insemination of stored semen and those born after natural mating. In the laboratory, there is a pressing need for long term storage of mammal spermatozoa, something which still cannot be done for the mouse or rabbit. If this hitch could be overcome, research could then be conducted on smaller animal colonies, with many of the males being replaced by banks of stored semen.

Storage of sperm is also an attractive idea for the conservationists. It is becoming impossible to maintain breeding nuclei of all species and all breeds. But, given the techniques of semen collection and storage and artificial insemination, all that would have to be conserved would be a restricted number of 'utility breeding stocks known to give fertile offspring when inseminated with the stored semen. By 'crossing' successive generations with one kind of stored semen, the breed or species of the utility stocks would gradually be transformed into that of the breed or species from which the semen came.

Techniques for long term storage of eggs and embryos of mammals have been nowhere near as successful as methods for storing semen. At best, only a few days' storage of a mammalian egg can be managed, and no one can yet say if, or when, the technique might be available for humans.

Efforts to culture eggs outside the body have also presented huge problems to the reproductive physiologists. Even in animals, we cannot as yet culture one and the same mammalian egg in the laboratory through all the successive stages of egg maturation, fertilization, early cell division and blastocyst formation—let alone implantation and development as a foetus. Much is known but the technical jig-saw puzzle has not yet been fitted together.

What can be done is to carry out various short parts of the sequence in a 'test tube'. I have already mentioned how immature ovarian eggs automatically start to mature when removed from an animal in the laboratory. The unfertilized rabbit egg, for instance, can be fertilized outside the body and will develop into an offspring, providing it is quickly transferred into a host mother. It is also possible to get rabbit eggs, already fertilized in the mother, to develop to the blastocyst stage in culture. Some workers have claimed to have cultured one-celled fertilized mouse eggs, but others have found difficulty in repeating such experiments. Cleaved mouse and rabbit eggs certainly develop to the blastocyst stage in culture, and there is one report of a one-celled fertilized cow egg which went through four successive cleavages in culture and became a 16-celled embryo. Though this work progresses with a great deal of optimism, these pilot experiments with animals

REPRODUCTIVE TECHNOLOGY in mammals is summarized in the scheme right, where the grey line connects various human processes and the black line connects processes of other mammals; broken lines connect processes which may yet become possible. The symbol S stands for long term storage, C for possible choice according to genetic contact and ± for contraception or abortion. In the male 'test tube' (top right) it is still only possible to store sperm, although in the distant future it may eventually become possible to produce animal sperm in the 'test tube' from a cultured testis. To date, in the female 'test tube' (bottom right) it has been possible to fertilize eggs of certain animals, and probably humans too; in other experiments, it has also proved possible to achieve the development of an early free embryo. In addition, unfertilized eggs have undergone a limited degree of parthenogenesis (virgin birth), subdividing a number of times to produce an early embryo. However, such embryos spontaneously 'abort' owing to their unfavourable genetic structure; a similarly limited process can occasionally take place in the female body. Nuclear cloning is another method postulated for by-passing the human fertilization process. Here the nucleus from a female body cell would be injected into an unfertilized egg devoid of its own nucleus, causing it to subdivide into an early embryo. So far, nuclear cloning in a 'test tube' has been achieved only with newts

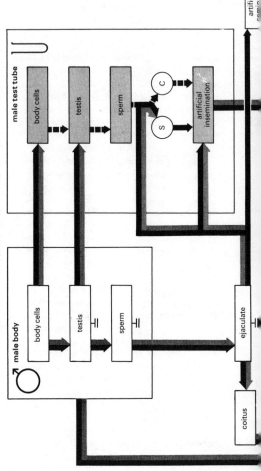

92

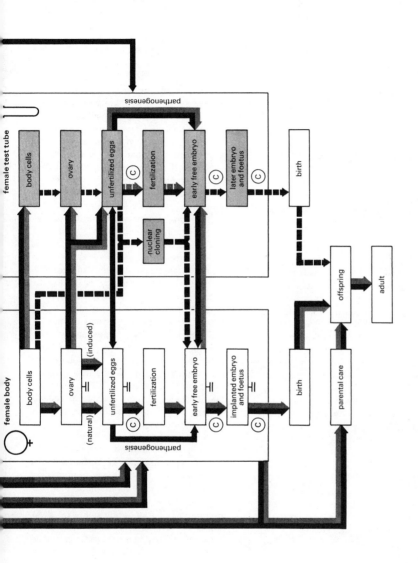

seem to give little hope as yet that an early or an easy solution to the culture of human eggs can be expected. But this takes us to a positive claim that is best considered in the next section.

FOR MANY YEARS there was no real success in trying to fertilize mammalian eggs outside the body. However, one definite success in the rabbit has been achieved: eggs fertilized *in vitro* and transferred back to a host mother have given rise to living young. In this kind of work, it is not enough merely to show that eggs cleave after being subjected to spermatozoa in a 'test tube'. All kinds of stimuli—including mere handling—are liable to provoke unfertilized mammalian eggs into an abortive parthenogenetic development which is difficult to distinguish from ordinary development.

Using, adapting and extending the work on animals by himself and others, Robert Edwards and his colleagues published a report last year on an experiment with *in vitro* fertilization in man. The eggs developed to a stage comparable with that attained in the first part of the normal fertilization process, though the state of many of the eggs was abnormal. There was provisional evidence that spermatozoa had actually entered the eggs. The publication gave no clear guide as to whether a provisional or a definitive claim was being made. However, the work received immediate acceptance from the world's Press. They may well have been right. Scientists in the field, however, have been a little more cautious; they have tended to regard this as fascinating work in progress, well worth publication and very likely to be on the right lines. But there are still many things to clear up, evident to the authors and to others.

The most advanced stage of development in all this work was that of a control egg described in a previous publication by Edwards and his colleagues; this was a two-celled parthenogenetic egg which had never been near a spermatozoon. Many of the putatively fertilized eggs had multiple nuclei within a cell, and this had certain consequences. First, the eggs bore a distinct resemblance to a series of parthenogenetic eggs which had not been fertilized. Second, the multiple nuclei of many eggs indicated a grossly abnormal distribution of chromosomes which would be expected to lead to early death of the embryo. But what of possible smaller and invisible chromosomal abnormalities which might allow an embryo to survive but be abnormal? Until this point is cleared up, there would seem to be uncertainties in human application. More recently, there has been stronger evidence that the spermatozoa actually enter the egg and that true fertilization occurs. One wishes all success to this venture in human biology; the implications for scientific and medical understanding are enormous.

Transferring of eggs and embryos from one female to another has been one of the more successful lines of research in reproductive physiology. It is easy to obtain living offspring after transferring eggs from one female to another in laboratory animals. Extensive studies in the mouse and rabbit have shown that the young are as normal as those born after normal processes of procreation. The technique is in routine use in many lines of research. (Transfer of spermatozoa is, of course, practised widely in a range of laboratory and domestic animals.)

The late Sir John Hammond, the reproductive physiologist from Cambridge, had a particular interest in the following possibility. The normal way of upgrading poor quality cattle is to mate them to high quality bulls through several generations. This unfortunately takes time. A desirable alternative, argued Hammond, would be to transfer into poor quality cattle the early embryos of cattle of higher quality, so that in a single generation all the offspring would be of high genetic quality.

Transfer can be carried out by surgical operation, but there is no economic future unless eggs can be transferred by less drastic means. Here a snag has developed. Eggs can be placed within the cow's genital tract by a simple non-operative technique (virtually the same simple apparatus as used for artificial insemination is employed to insert fertilized eggs instead of spermatozoa). But however carefully one carries out the sterile precautions, a protective mucus seal at the entrance to the womb has to be pierced and gross infection of the uterus commonly results. Non-operative transfer has been carried out in mice, goats and pigs, but the pregnancy rate is low.

One unusual and specialized aspect of egg transfer may yet have some economic value. It is based on the fact that sheep eggs remain viable for several days when transferred into a rabbit. This fact recently permitted a successful experiment in which sheep eggs in Britain were transferred into rabbits which were then sent by air to South Africa. There the eggs were recovered and transferred into host sheep which later produced lambs. In effect, several sheep had been transported in a package whose net contents weighed only a kilogramme or so. But for all the success achieved in animals, a practical technique of egg transfer is not yet available for use in human beings. From the work with animals the indications are that surgical transfer might be possible—but difficulties might be encountered in developing the much more desirable technique of non-operative transfer.

UNTIL THE MOMENT of birth, we normally have no means of telling what sort of baby will arrive and no means of ordering the kind desired. In specialized cases, usually when events are thought to be going wrong, something can be told about the nature of the foetus by use of X-ray equipment, or from prior knowledge of the parents genetic status.

Many scientists are certainly alive to the possibility of direct monitoring of the genetic status of foetuses, embryos and even eggs and spermatozoa. Given various ethical considerations, medical intervention can encourage or discourage the birth of children of certain genetic types—specifically, for instance, to determine whether the child is male or female. In mammals, in general half of the spermatozoa contain a Y chromosome and produce boys; the remainder carry an X chromosome and produce girls. Normally the two types fertilize in a proportion close to 50:50. Can this proportion be altered experimentally, so that the sex ratio is controlled before conception? Positive claims made for control of the sex ratio always receive immense publicity, while justified criticisms or failure to repeat results usually go unmentioned in the Press (one can scarcely find a more striking example of the mass media's lack of concern for a balanced account). There is not the room here to present all the details; I will simply record some personal conclusions which I have justified elsewhere.

Interesting positive results have been reported in the scientific literature, but I am *not* yet convinced by any of the following claims: that X and Y bearing spermatozoa are a distinctively different shape under the microscope; that they have been separated from one another by experimental means; and that the sex ratio in any mammal has been changed by experimental intervention before conception. There are strong theoretical reasons for believing that X and Y bearing spermatozoa cannot differ in any way over and above the very small physical difference caused by the fact that X and Y chromosomes are certainly not identical. Perhaps a separation of the two kinds can be based on this small difference.

However, once fertilization has taken place a technique does exist for monitoring the genetic status of eggs, embryos and foetuses. In the rabbit, for instance, it has been shown conclusively that the sex of embryos can be recognized at a very early stage when they contain only a few hundred cells. Here the technique used by Gardner and Edwards has proved to be direct and simple. The blastocysts recovered from a rabbit are pricked, causing a little blob of cells to be extruded which can then be cut off with a pair of scissors. A quick microscope preparation of the blob enables it to be sexed by inspection of the 'sex-chromatin'—a small dot of chromatin seen in most female cells but not in male cells. The blastocyst itself is cultured until the wound heals and, after transfer into a recipient rabbit doe, it produces an offspring of the anticipated sex!

The experiments carried out by Gardner and Edwards were on a sufficient scale to prove the efficacy of prediction. This work was confined to the recognition of sex, but it paved the way for early recognition of certain anomalies in which unusual numbers of sex chromosomes are present. In human beings, for instance, embryos with anomalies in the number of sex chromosomes, or in the number of ordinary chromosomes, develop into individuals with specific mental and physical defects.

Still later in development, several kinds of genetic status can be recognized without handling the embryo or foetus in any way. Around the developing foetus is a cushion of fluid, the amniotic fluid, into which embryo cells wander. Some of these embryo cells can be collected simply by sucking out a small amount of fluid and then they can be multiplied in tissue culture. By examination of their chromosomes and sex, chromatin, it is possible to say whether the foetus is male or female and whether it is afflicted by one of a variety of chromosomal abnormalities known to give rise to malformed or mentally deficient children and adults. There is some slight risk even with this simple technique and for this reason, and because we cannot afford to dissipate medical resources, the medical profession is unlikely to apply the technique to apparently normal pregnancies, where prediction of sex at this stage does little more than enable one to decide a few months earlier than usual whether the baby's clothes should be blue or pink!

Prediction or control of the sex of the unborn also has a direct connection with the wider question of predicting other aspects of the genetic status of the foetus. There are several well known instances of 'sex linked inheritance' of congenital defects caused by genes carried on the X chromosome. In other words, the same chromosome that determines sex can also carry a deleterious factor. For instance, the fairly severe human disease called haemophilia, in which the sufferer is subject to massive bleeding even from minor wounds, is carried in this way. Occasions arise when the genetic status of a mother is such that she has a 50 per cent chance that any of her sons will suffer from haemophilia (there is no danger of her producing a haemophiliac daughter, since female haemophiliacs are to all intents and purposes unknown). If such a woman is not to have a haemophiliac offspring, then she must not bear sons—and this can be arranged by medical intervention. So, in short, it is already possible in human beings as well as animals to recognize the sex and certain other genetic properties of the embryo or foetus, some of which may call for a termination of the pregnancy. But control of sex before conception is still an unknown matter for the future.

WHAT, HOWEVER, are the chances of actually by-passing fertilization? Can an embryo be produced from an unfertilized egg?

In newts, which are lowly animals but nevertheless vertebrates like ourselves, a fascinating technique called nuclear cloning has made it possible to produce a 'clone' of large numbers of genetically identical animals—the genetic equivalent of multiple identical twins. This is done by destroying the nucleus in a number of eggs and injecting into each egg a replacement nucleus from a later embryo. As far as the egg is concerned, the fertilization process has then been by-passed. The important point is that the substitute nuclei can be taken from the numerous genetically identical cells of a single embryo. The eggs with their substitute nuclei then grow up into genetically identical newts. It seems likely that in the near future we can expect

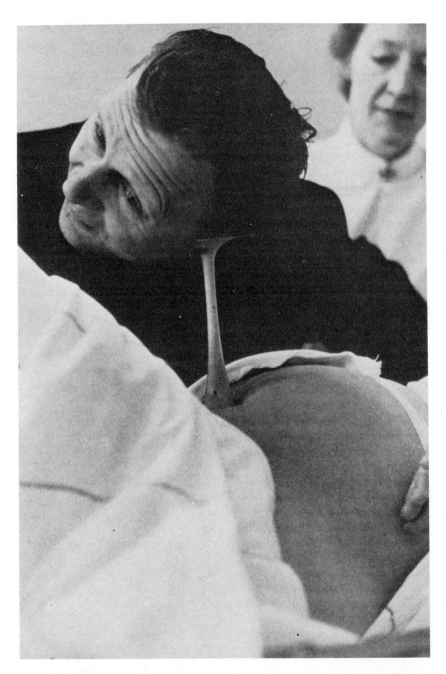

HEART BEAT of the unborn child can be readily heard by the fourth month of pregnancy; embryonic heart beats have in fact been detected as early as the eleventh week. Normally there is no means of telling the sex of the baby until the moment of birth. But a research technique is now available in which embryo cells found in the amniotic fluid are sucked out and multiplied in tissue culture. By examining their sex chromatin it is possible to say whether the foetus is male or female, or whether certain chromosomal abnormalities exist. There is, however, some risk with the technique and it is reserved for cases where the genetic status of the mother suggests a possibility of the offspring inheriting a congenital defect

97

experiments of this kind will at least be attempted in mammals, and that substitute nuclei may be taken from adult, rather than embryo, tissue.

Another phenomenon well known in lower animals—parthenogenesis or virgin birth—by-passes fertilization altogether. In this case the egg simply develops, without a spermatozoon, into an embryo and adult. This is the normal course of reproduction in certain invertebrates such as the greenfly and in some species of fish and reptiles; and has been achieved experimentally in amphibia and in the turkey. Large numbers of parthenogenetic blastocysts have been produced in laboratory mammals such as the rabbit, but none of them ever seems to come to term as a born offspring. There appears to be a block to full development of parthenogenetic mammalian embryos, and no one knows if, or how, it can be overcome. In humans, it can be said fairly definitely that parthenogenesis never occurs, not even as a rare spontaneous event; thousands of women have been kept in strict isolation in penal and religious establishments, and there is no authenticated record of any inexplicable conception under such circumstances (save in terms of human fallibility and a hitch in the administrative machinery of the establishment).

In any case, parthenogenesis is probably not a very desirable process for humans. What we know of lower vertebrates tells us that there are two chief kinds of parthenogenesis—one of which produces the equivalent of a high degree of inbreeding, the other the equivalent of 100 per cent inbreeding. Unless a sudden and unexpected breakthrough is achieved, neither nuclear cloning nor parthenogenesis seem to be likely possibilities in man in the near future. The techniques are certainly not with us at present, and no application can be envisaged save in a science fiction world. In formulating our ethical attitudes to such possibilities we should remember that 'cloning' is not in itself an 'unnatural' state of affairs: ordinary identical twins are a miniature clone, and the armadillo regularly produces identical quadruplets.

A PAINFUL DICHOTOMY has grown up between scientists and non-scientists concerning the application of reproductive technology to man. It is part of the same general misunderstanding between the two cultures—to borrow Sir Charles Snow's phrase—which has become so apparent with the current general threat to our environment. One particular aspect is the 'naturalistic fantasy' world in which such a surprising number of people seem to dwell. Here is a world where change is resisted: where crops must be manured with natural animal faeces, the equivalent chemicals being artificial; where there are two kinds of fluorine, natural fluorine in the soil and water, and unnatural fluorine—the same element—which is added to correct a natural deficiency; where one's own blood is natural and transfused blood, or antisera from the blood of an animal, is unnatural. In this world, any aspect of reproductive technology is especially frowned upon.

The use of reproductive technology in man means some degree of change —however small—in our reproductive habits. The very structure of our language makes it difficult to discuss such changes objectively; the undertones of meaning in a word often favour the continued existence of the culture group in which the word is used. For instance, there is a kind of crazy alliterative connection—a perjorative rhyming jingle almost—connecting the following words: adultery, adulterate, mass medicate, miscegenate, substitute, prostitute, prosecute. Among these emotive words are several which are frequently used by opponents of change of one kind or another.

However, those who oppose change in our reproductive habits and processes do have one very important criticism. They argue, sensibly, that if new forms of control of human reproduction exist, or are likely to be developed, who is to control the controllers? Let us not evade the issue: many discoveries have been, and will continue to be, misused. (Should the cure for malaria have been withheld—because it has led to over-population? Should the development of penicillin have been stopped —because it kept alive soldiers fighting unjust wars?) Of course it is no answer to call for an end to all research so that there are fewer discoveries to abuse.

The avoidance of the birth of children with clear-cut abnormalities—negative eugenics—is already fairly acceptable. But scientists and non-scientists alike have a powerful aversion to positive eugenics—the controlled breeding of 'better' children. The argument is threefold. First, wherever it has been attempted on a large scale it has been in the hands of evil men. Secondly, there is no proper measure of indefinable qualities such as nobility or courage; and if we cannot measure a character, we cannot select for it. Thirdly, even if selection were effective, the results would not necessarily be something to look forward to with pleasure. The controllers of positive eugenics would probably be leaders of religious or political power groups. Desirable human qualities—those tending to perpetuate the power group—would no doubt include submissiveness to the power group and readiness to act rashly on its behalf. Undesirable qualities would probably include gentleness and the questioning of the power group's authority. With relief, one must conclude that mankind is simply not ready for positive eugenics.

But although we are generally conservative in our attitudes, the human race does have a healthy capacity to undergo slow evolution in its reproductive habits. In the Muslim world, for instance, simultaneous polygamy was once the norm but the people are now tending towards monogamy. By contrast the monogamous system of the earlier American settlers has swung towards sequential polygamy—now on its way to being the norm for the western world.

Those who find the whole idea of controlling our reproductive habits unthinkable would do well to consider the extremely detailed control we already accept. It is much like the control exerted in breeding pedigree animals. Matings must not take place before a certain age. The mating couple must be registered, as also the offspring. A change of partners must be registered. And there are social and official penalties for the unregistered.

If, at the individual level, the problem is to reduce fertility, then contraception technology is well advanced in limitation and planning of family size. If the problem is to promote fertility, surgical or hormonal treatment can remedy anatomical defects or correct chemical deficiencies in women. If there is a block to normal fertilization, it may yet be possible to produce children after no more than a brief sojourn of the gametes outside the body, depending on the progress of current research on *in vitro* fertilization. If the infertility is due to an inability to bring viable spermatozoa to the unfertilized egg by the process of normal coitus, then artificial insemination can sometimes be of use. If a husband is sterile, there is always artificial insemination by donor (AID) to give a woman a child. If a woman cannot produce viable eggs of her own, she may (if current research reaches fruition) be able to bear a uterine foster child—effectively adopting a child *before* instead of after birth.

Then there is the clear-cut case of women who have a definite chance of producing an abnormal child such as a certain type of Mongol. In the past, the cure for this has been to have no children, which has meant contraception or sexual abstinence. But an alternative method is now practical: monitoring the qualities of the unborn child to allow it to be born if normal, but to terminate the pregnancy if abnormal. If, and when, methods of culture and transfer of fertilized eggs mature, it would then become possible to identify the abnormal embryo—and terminate the undesired pregnancy—at a much earlier stage before anything resembling a foetus has developed. Here, I must emphasize that reproductive technology of the kind envisaged in this article cannot eliminate deleterious genes from a population. If you could prevent the birth of all children suffering from a given genetic defect, then a single generation might indeed be produced in which no one displayed the defect. But you would not have controlled the natural mutation rate; nor would you have eliminated the carriers of a recessive gene who, while showing no symptom themselves, could pass on the genetic defect to a future generation.

The problems of infertility, important enough to the individual, are not at present population problems. They might become so, in which case half the problem is already solved by our ability to store human semen outside the body. The other half—storage of fertilized or unfertilized eggs for transfer into women—is not yet feasible. The complete solution—safe storage of spermatozoa and eggs with subsequent development *in vitro*—is still in the realms of science fiction.

It is, however, fertility and not infertility which is the current problem of the world. The elimination of the 'unfit' by starvation or disease has become ethically unacceptable. And so the population multiplies. Even the fallible, though infallibly recommended, rhythm method of contraception and the more reliable 'pill' are both little more than useless in the underdeveloped world since they require a level of education and prosperity which currently does not exist. The only effective method of population control in such countries would be by imposing various contraception methods without the consent of the inhabitants. And as one of the few prophecies in this article, I forecast that this would almost certainly be conducted by the wrong people for the wrong motives.

At present, reproductive technology is applied to human beings for the control of fertility and infertility and is on the point of being able to prevent abnormal children coming into the world. The methods will improve as the diagram on page 95 becomes filled out by research. Anyone who lets his imagination loose on the diagram will be able to visualize other, more startling, applications which could become profoundly important if the human race were ever to find itself in some unforeseeable genetic emergency. In the meantime, and for a long time to come, the union of man and woman will produce children in the usual way.

FURTHER READING
THE FUTURE OF MAN by P. B. Medawar
(Reith Lectures, Methuen, London, 1959)
A RUNAWAY WORLD? by E. Leach *(Reith Lectures, BBC, London, 1967)*
SEX, SCIENCE AND SOCIETY by A. S. Parkes *(Oriel Press, Newcastle, 1966)*

UTOPIAN MOTHERHOOD AND
POPULATION CONTROL

By Dr. Robert T. Francoeur
**Associate Professor of
Experimental Embryology at
Fairleigh Dickinson University**

The specter and the creative challenge of the man-made man, reproductive technology, and "utopian motherhood" are with us today.

Yet despite our concern with a mushrooming populace in an ever more polluted and depleted environment, few people, even among the more educated, have given serious thought to the potent applications and repercussions which new trends in human reproductive technology can have in dampening and perhaps even halting the population explosion.

Discussions of the population-environmental crisis almost invariably stir a hopeful exploration of the latest contraceptives and related mass education programs as if this were the sum and substance of reproduction technology. From a perspective in the field of experimental embryology and its social impact, I am convinced that this approach is shortsighted.

Today, particularly with our near instant worldwide communications, a new and revolutionary technology does not have to be accepted and used extensively before it begins to exert a major influence in shaping society's behavior and thought patterns.

Only a few men and women have benefited from a heart transplant operation, yet this technique has already radically modified our legal and common understanding of death. The psychological and legal repercussions of our ability to transplant the human heart have clearly preceeded its extensive use. In fact, there is a good likelihood that this impact will continue even if the operation does not become commonplace.

CATALYST FOR ENVIRONMENTAL QUALITY, 1971, Vol. 1, pp. 7-10.

Similar examples abound in the fields of reproduction technology and experimental embryology. And some of these, to my mind, give us much greater hope for a solution to the population crisis than a simple focusing on the practicalities of contraceptives.

Around the turn of this century two British scientists, Dr. Walter Heape and F. H. A. Marshall, tried to transplant some cow embryos from their natural mothers to surrogate or substitute mothers. Fifty years passed before Raymond Umbaugh succeeded with the first artificial inovulation, as the transplanting of embryos has since been termed. By the late 1960s, embryo transplants had become a common practice in the breeding of cattle and sheep.

Scientists today can chemically induce superovulation in prize ewes and cows and collect the eggs for test tube fertilization or use artificial insemination. After incubating the resultant very early embryos in a rabbit surrogate mother for a few days, during which the prize embryos can be shipped by air freight almost anywhere in the world, the offspring can be transferred to their genetically inferior but sturdy surrogate mothers' wombs for "normal" pregnancy and delivery. Shipping a rabbit incubating a hundred prize calves is far easier and cheaper than trying to produce the same offspring the normal way, or even with artificial insemination.

Artificial insemination and embryo transplants may well prove to be our most valuable weapon in preventing the disappearance of many endangered species of wildlife.

Twenty species of mammals are today on the verge of becoming tomorrow's fossils. Less than 4,000 tigers roam the jungles of India, and more orang-outangs live in zoos than are free in their native Borneo.

But what if we were to apply artificial insemination and embryo transplants to this problem? Sheltering wildlife in parks and reserves often reduces breeding. These same game preserves, however, could provide the necessary animals for some promising experiments.

For instance, the European bison, down to less than 100 after World War II, could be superovulated and either artificially inseminated or test-tube fertilized. The resultant zygotes could then be incubated in rabbit uteri for shipment to our western states where artificial inovulation could be arranged, with the plentiful American bison serving as substitute mothers for the pregnancies. Similarly, the British and Russians could superovulate and inseminate their precious but cantankerous pandas which have refused to reproduce the normal way.

In the spring of 1970, a 35-year-old housewife from Lincolnshire, England appeared on a BBC television program to announce that she hoped to have a child soon, after seven years of childless marriage. Mrs. A then asked the British public to contribute financial support for the work of Dr. Robert Edwards and Dr. Patrick Steptoe in achieving the first human embryo transplant.

Both of Mrs. A's fallopian tubes are blocked, but it is hoped she can bear a child of her own in the following fashion.

Dr. Edwards will collect Mrs. A's superovulated eggs surgically, fertilize them in a test tube with her husband's semen, perhaps incubate them in a rabbit womb for a few days, and then try to get her womb to accept one of the zygotes for implantation and a normal pregnancy.

Forty-eight other British women have joined Mrs. A in seeking a solution to their infertility by such inovulation technology. That they do not resort to annonymous or known surrogate mothers to carry their children hardly lessens the psychological impact of this startling new way of producing a human being.

Unlike the earliest cases of human artificial insemination in the 19th century, this medical advance is not hidden under the silence of some Victorian blanket. Instead Mrs. A and her husband share their emotions and yearnings with millions of television devotees in countless pubs, schools, and homes within the reach of the British Broadcasting Corporation. Newspaper accounts further extend the psychological impact around the world.

The legal, emotional, moral, and social complications we can expect to face in the near future—when embryo transplants and surrogate mothers become as much a reality as heart transplants, or more so—do not, in my mind, detract at all from the psychological reinforcement this technique lends to our broad hopes of successful population control.

Before I explore that reinforcement, let me detail some other developments in human reproduction which, I am

103

convinced, indicate *a major psychological revolution* emerging rapidly among men and women in this country and elsewhere.

The use of human semen to bypass sexual intercourse and still produce a normal pregnancy and healthy child, can be traced back to 1799 and an English physician, one Dr. Home. During the American Civil War, Dr. J. Marion Sims attempted, with very modest success, what he called "ethereal copulation." And in 1909, Dr. Addison Hard Davis brought the technique of artificial insemination to public attention with an article in "Medical World."

Conservative estimates already suggest that over a million Americans cannot attribute their origins to impregnation in the orthodox manner. More than 20,000 Americans are starting life each year without benefit of sexual intercourse. And this number is rising rapidly as genetic clinics and advances in human genetics make eugenics a more common reason than sterility for artificial insemination.

At the next Cannes Film Festival the documentary section is likely to show a rather unusual 10-minute film. Even in its roughest, unedited form—which I viewed recently—the film is strangely fascinating and disturbing. It shows lamb fetuses, removed prematurely from the womb, floating in plastic, aquarium-like "wombs" with their umbilical cords connected to an array of instruments: extraembryonic membrane oxygenators, two tanks of oxygen and a pump to replace the mother's heart, and a dialyzer in place of her kidneys. As the oxygen level is shifted in the blood supply, the fetus begins to suck on a scientist's finger in the "womb." It becomes increasingly restless and soon is ready for "decantation" birth.

Complementing this work of Drs. Warren M. Zapol, Theodor Kolobow, and Gerald G. Vurek at the National Heart and Lung Institute in Bethesda, Maryland, are dozens of other scientists around the world working on a variety of artificial "wombs."

While the "telegenesis" of artificial insemination with fresh or frozen semen has separated sexual intercourse from procreation in a very practical way, the psychological impact of artificial womb pregnancy is even stronger.

As with artificial inovulation, the practical application of ectogenesis is not yet a reality. But the artificial womb may be available in the near future to help save the more than

104

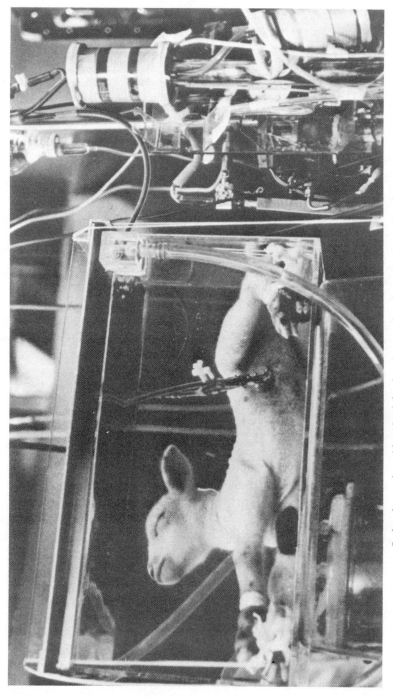

Isolated, non-breathing fetal lamb submerged and resting in tank of synthetic amniotic fluid in the National Heart and Lung Institute's Laboratory of Technical Development. The lamb is being provided total respiratory support by an "artificial placenta" system developed in the laboratory.

25,000 premature babies who die in this country every year because of a hyaline membrane disease. This application would make the likelihood of a full term ectogenic pregnancy even more probable in the not too distant future.

The psychological impact of artificial insemination, artificial wombs, and embryo transplants, however, is minor when compared with the possibility of completely bypassing all the sexual elements of egg, sperm, intercourse, and fertilization. This possibility, technically known as "cloning" or asexual reproduction, may never become anything more than a laboratory technique for exploring the intricacies of human genetics and the age-old controversy of heredity versus environment. But it carries to ultimate and logical conclusion the psychological and practical separation of sexual intercourse and procreation.

The work of Dr. John B. Gurdon and Dr. Ronald Laskey has, within recent months, brought closer to reality the possibility of one day producing a human being from a single isolated adult skin cell. Their work involves growing isolated skin nuclei from adult frogs in tissue culture and then implanting these specialized adult nuclei in ennucleated eggs where somehow they return to their embryonic state and produce a frog without benefit of sexual intercourse.

Joshua Lederberg, Nobel Prize geneticist at Stanford University, has predicted that within 10 or 15 years he will use an adult human skin cell to completely bypass sexual intercourse, sperm and egg, and produce the first cloned member of the human race—the identical twin of his "progenitor" except that it will be 20-40 years younger.

Before mankind can morally, psychologically and emotionally accept any program for limiting the population, we have to experience and accept a major and fundamental shift in our thinking on human sexuality.

For thousands of years mankind has lived by a more or less clear equation between sexual intercourse and procreation. While the scientific evidence and understanding was not available to support this connection, it was accepted as part of the natural law—the way the Creator had established things in the beginning.

This equation was further reinforced in a variety of religious traditions which reflected the fact that religious leaders found it practically impossible to incorporate the passionate character of human sexuality and intercourse into the context of man's "spiritual" or "religious" life.

To rule out the possibility of sexual intercourse and deny human sexuality has never been a popular solution. Thus most religious traditions, including those within the Judaeo-Christian current, have accommodated human sexuality and intercourse by first tolerating and then exalting its practice within marriage for the sole purpose of procreation.

In the past half century, however, there has been a move on the part of some Christian groups to accept sexual intercourse as an interpersonal experience whose value is distinct and separable from its procreative possibilities. Many factors—social, cultural, political, and technological—flow together in triggering this shift. Certainly the social emergence of women, and a growing awareness among Christians of their painful indebtedness to pagan dualistic thinking, are prime among the agents.

The potentialities of artificial insemination, test tube fertilizations, artificial wombs, embryo transplants, asexual reproduction, and prenatal monitoring with increased survival rates may seem to pose a serious threat to solving the population problem. My appraisal, however, is that they are a potent ally if we use their psychological impact to reinforce that of "the pill." These techniques force the undeniable conclusion that sexual intercourse is no longer linked with procreation, and that for mankind today the two are and should be distinct human acts, each governed by its distinct ethics.

If we utilize this message properly, we can deal far more effectively with the population problem than if we simply make contraceptives universally available to try to persuade people to use them for the sake of controlling population.

For instance, if we alloted the time, effort, and money needed to develop both the technique and the facilities for long term cold storage of semen (and/or eggs) without genetic deterioration, artificial insemination could provide one of the most potent controls of our expanding population.

One way this might work is as follows. Shortly after reaching puberty each and every young girl would visit her family doctor for a thorough checkup. If everything proved normal, the doctor could give her an injection of hormones combined from pregnant women and pregnant mares. These would shock the girl's physiological system so that instead of releasing a single egg in the middle of her next monthly cycle she would superovulate several dozen eggs. As they passed down her fallopian tubes, on their way to the womb, these eggs could then be collected with a simple syringe technique for storage in a deepfreeze. The container would, of course, be labeled as hers and officially sealed. The girl then could have an operation, similar to an appendectomy, in which her fallopian tubes were cut and tied. Afterwards she would have a normal monthly cycle with normal ovulation, but since her egg could no longer come together with a man's sperm, conception would be impossible.

Each and every teen-age boy might follow a similar procedure, without the hormone injection, and have a large sample of his semen collected for storage. Unlike the following operation in the female, which involves complete anesthesia and major surgery, the cutting and tying of his vasdeferens (which carry the sperm out of the testes) is a minor operation involving only a visit to a doctor's office and a local anesthetic.

Some years later, one of these girls and one of these boys would meet, fall in love, and marry. When they decided to enrich their marriage with a child, they could visit the dispensaries where their eggs and/or semen, were stored. If only the husband's semen was stored and the wife had not had an operation, simple artificial insemination could be used. If only the wife's eggs were stored, the doctor could defrost one of these and implant it back into her tube for fertilization in the course of normal intercourse. If both germ cells were frozen, test-tube fertilization could be followed by implanting the fertilized egg in the wife's womb.

This approach has many promising aspects and very few disadvantages. Our prime advantage is the bypassing of all the practical complications, side-effects, and unknown long-range effects of today's contraceptive pills and of the IUDs. Several state health department surveys have indicated that as many as one out of every six girls between the ages of 13 and 19 is illegitimately pregnant. Puberty sterilization would eliminate all unwanted pregnancies while providing a means of directly and effectively limiting the number of children allowed each married couple.

In the near future governments may find it necessary to sign population limitation treaties even as they now agree to arms limitation. Dr. Linus Pauling and Dr. Robert Shockley,

Nobel Prize-winning scientists, have both suggested sterilization at puberty as one certain technique for population control. Interestingly enough, the opposition of moralists and religious leaders to this approach is not as vehement as one might expect. Jesuit theologian Thomas Wassmer, for instance, expresses the reluctance most people have to such Federal control, but he sees it as morally acceptable if necessary.

Ashley Montagu and other sociologists have argued that the contraceptive pill ranks among the dozen or so major revolutions in human history. I would place "the pill," and the associated technologies of reproduction discussed above, at the very top of that list.

Other revolutions—the wheel, the discovery of fire, the invention of gun powder, the steam engine, nuclear power, electronics, jet transportation—have each had their impact on society. But the technologies of utopian motherhood go far beyond the externals of our lives touched by these other major revolutions.

Reproduction technology cuts to the very quick of every human being. It forces us to ask some very basic and crucial questions about our role-image of man and woman, our concept of family and parenthood, our appreciation of human sexuality.

Today's reproductive technology forces every man and every woman who learns of it to accept, whether or not we like it, the psychological distinction between two now separate human actions—sexual intercourse and reproduction. Whereas in the past we have lumped the two together and morally justified the one by the other, now we must face the task of evolving and articulating *two totally distinct ethics*. There must be an ethic for procreation, which takes into consideration the concrete realities of an exploding population. And there must be another ethic governing sexual intercourse, more likely based on interpersonal values and responsibilities.

(Ed. Note: For further details of the revolution now going on in the field of experimental embryology, and the new trends in human reproduction likely to result, we refer you to Dr. Francoeur's recent book, "Utopian Motherhood," published by Doubleday.)

The Human Multitude: How Many Is Enough?

By BENTLEY GLASS

The question "Is there an optimal size of the human population?" poses to a geneticist the immediate need for definition of several terms. First, of course, one must ask: optimal for what? Second, precisely what are the definitions of "population" and "size of population"? Third, is one to be concerned chiefly with the welfare of man in the immediate present or in the distant future? Finally, if the optimal size of population is related to changing parameters of the environment, can we expect to preserve an environment sufficiently buffered against change that a prescription today will have any validity whatsoever for tomorrow?

Conditions of Human Adaptability

A biologist must examine these questions in the light of "mankind evolving," to use Theodosius Dobzhansky's apt phrase. The primary characteristic of *Homo sapiens,* in contradistinction to other mammalian species and in particular his closer primate relatives, is the high order of intelligence he possesses—a quality that enables him to adapt his environment increasingly to his own desires instead of awaiting the far slower genetic adaptation of himself to his changing environment. Man's own adaptability, including his power to learn from experience and, through science and technology, to enlarge his power to exploit and modify the environment, depends on the considerable diversity of the

THE AMERICAN BIOLOGY TEACHER, 1971, Vol. 33, pp. 265-269.

human gene pool. It seems to me that although a "brave new world" is conceivable—that is, one in which man is artificially constrained to become genetically homogeneous or is reshaped into a few genetic castes—it is neither likely to develop within a democratic society nor is as optimal as a continuance of extensive genetic diversity. For mankind, a continuance of his adaptability and maintenance of the diversity of his gene pool would seem to be optimal, at least for the remote future.

According to Sewall Wright (1960), the situation most favorable for progressive evolution is one that is provided by an oscillation between a relatively large population homogeneous in nature and a state of subdivision into many local, fairly isolated populations, at least some of which are of medium or small size. In that case the small isolates will come to differ from one another both in concealed genetic variability (genotypes) and in visible characteristics (phenotypes), and natural selection may discriminate between them. This view is generally accepted by population geneticists. Only in relatively small and isolated populations, because of inbreeding, do we find the genetic variability forced into expression and thus made subject to selection. On the other hand, if the total population of the species consists of only a few small isolates, there is too much danger that none of them can provide a phenotype permitting preservation in the struggle for existence. Here both chance and adverse selection will combine to magnify the likelihood of extinction.

In the one or two million years of existence of the human species, the total population was at first undoubtedly small, and it was also broken up into small, local breeding units that might be designated as bands, or extended families. Inbreeding was high; diversification among the isolates was probably great. The risk of extinction of the species was overwhelming until the total population of the species rose to 100,000 or so, when there would have been many bands. Later, as group methods of hunting developed and the usefulness of fire as a protective weapon was discovered, larger tribes succeeded the small bands. Ever since, we have moved steadily toward larger and larger Mendelian populations, with always a higher proportion of the variability in the gene pool concealed, and with less and less diversity of isolates and racial groupings. The human melting-pot came into being. The melting-pot has not yet eliminated all racial distinctions, but it is rapidly doing so. Genetic variability now chiefly differentiates individuals of the same breeding population, and a greater proportion of it is concealed in the recessive state. Populations have grown so vast that sheer distances, both geographic and social, have become primary factors in offsetting a purely random breeding pattern. Evolutionary change has thus slowed down, even without considering the diminution of the rigor of natural selection that has also taken place because of better standards of health and nutrition throughout the world. As the population structure of man has thus changed, his evolutionary prospects have altered.

Faster Change, Slower Selection

The present rate of change in the human condition is exceedingly great. No genetic change can occur fast enough to cope with it even under conditions of maximum mutation rates and rigorous selection. Mutation rates are in general low, although the great size of present human populations ensures that many mutations, of each and every gene; will occur in every generation. The new genes will also undergo extensive recombination, so that many different coadapted gene systems will exist. At the same time, our large populations ensure that most of the genetic variability will be concealed in the recessive state. Selection will consequently act mainly on the heterozygous effects of genes, which are commonly smaller in extent than the homozygous effects. Selection has been further slowed down by the advances in medicine and nutrition that now virtually guarantee that every child born alive, and some that previously could not have been born alive, will reach the reproductive years. The reduced rigor of selection is increasing the abundance of once-detrimental genes in our population, for these genes no longer lower their possessor's chance of reproduction. Dominant genes that are reduced from $s = 0.5$ (elimination of half the genes in each generation) to $s = 0.01$ will tend to increase 50 times in frequency. Recessives undergoing a similar moderation of selection will increase 10 times in frequency. Thus, in anomalous fashion, we have found a way to increase our genetic variability without requiring any alteration of the mutation rates—though these, too, may become greater through greater exposure in the modern world to ionizing radiations, chemical mutagens, and increased temperatures.

This genetic variability, however, is of little advantage to us in producing any real adaptation. We now modify the environment around us by technological means and create for ourselves a novel and artificial world in which our defective genes, even though active, fail to impede us in reproduction. The man of tomorrow will clearly need many more pills and prosthetic devices. That need not worry us so long as the social cost is not too great and so long as we can maintain our artificial environment unimpaired. By natural standards we may become degenerate physical beings, but we will not wish to apply natural standards any longer.

What is the optimal size of population? Man has achieved his present status chiefly through intelligence and adaptability. That evolutionary process occurred in populations very much smaller, and much more broken up into breeding isolates, than we now see. Evidently the great size of our present populations is not needed for progressive evolution and the maintenance of diversity. Our new condition provides us with added genetic diversity, but at a considerable cost: the social repair of genetic defects. This excess of diversity seems neither necessary nor advisable.

There is, I believe, only one condition under which it may actually save our species. If in the end we blunder into a nuclear war and thereby eliminate nine-tenths of the human lives on our planet, the scattered survivors will find themselves once more in the population structure of early Pleistocene man. Those who survive the direct effects of heat and blast and radiation and who are later spared by the fallout may still perish in large numbers because of their inability to provide themselves with the special foods, drugs, and devices on which they depend for life itself. Among the others, in remote inbreeding isolates, there will very likely be sufficient genetic diversity to enable *Homo sapiens* to make a wiser fresh start.

Improving on Nature's Ways

It should be feasible, by the year 2000, to bank human reproductive cells of both sexes in frozen state, as we now do with the sperms of domestic animals, especially sheep and cattle. In this way the reproductive cells of selected individuals might be utilized even long after their deaths to produce in the laboratory embryos that might be implanted in the womb of a foster-mother, or might even, after sufficient development of technique, be grown in bottle cultures. The latter "brave new world" technique I do not expect to see realized by the turn of the century. On the other hand, I do expect that techniques will be developed for the cultivation in the laboratory of portions of human ovary and testis, permitting successful continuous production of mature ova or sperms. Recent successes in the production of mature ova from cultured mouse ovaries lead me to expect that only persistence by a sufficient number of skilled biologists is needed to attain successful cultivation of human reproductive organs, continuous production of eggs and sperms, and formation by fertilization in the laboratory of as many human embryos as may be wished.

I am frequently asked why anyone should wish to pursue this goal. "Aren't the age-old ways of making babies good enough?" Several reasons may be given why such exploration of new possibilities is desirable. Only by studying the development of the human embryo and fetus under continuous observation and under various conditions can medical scientists really learn what factors produce particular kinds of abnormalities and how these may be corrected or avoided. Moreover, the practice of "prenatal adoption"—that is, the implantation of a healthy selected embryo in a foster-mother's womb—appears to meet with fewer religious and legal objections than the present practice of artificial insemination of a woman, whether with or without the consent of her husband. The development of the implanted fetus within the mother and its normal delivery at full term will engender far stronger maternal and paternal feelings in the "parents" than adoption of a child already several years old. Moreover, most couples who are sterile may in this way have virtually all the experiences of parenthood.

There are other reasons why such practices might be adopted—if not in the United States, then possibly in other countries. Banking of reproductive cells taken from individuals around 20 years of age may serve to prevent the accumulation of detrimental mutations with advancing age. From the standpoint of eugenics, parents should have their children while they are young. From the standpoint of economics and population control, marriage is often long postponed, and children are produced by parents in their thirties and forties. Healthy children may of course be produced by parents even at an advanced age, but the odds are not so good. By banking reproductive cells under conditions where mutation is reduced to the lowest level, and using implantation of an embryo produced by artificial fertilization, older parents may have children as free of defect, on the average, as when they were young. Persons such as astronauts who venture into especially hazardous environments, where they are very likely to be exposed to sizable doses of high-energy radiations, might be similarly well-advised to bank their reproductive cells under safe conditions.

These practices would of course be ineffective without employment of effective contraception. In spite of personal and religious objections, it seems clear that the use of present means, especially of steroid "pills" and intrauterine loops, will soon become worldwide. A promising recent advance, still being tested for safety and efficacy, appears to be that of inserting a dose of progesterone, enclosed in a capsule, under the skin of the female. Under these conditions, microdoses seep into the circulation and prevent conception even without modifying the usual female cycle. The capsule can be easily and painlessly put in place and may be removed at any time. Modern methods of contraception, it seems clear, will if adopted be the quickest and surest means of securing control over the population increase; and they seem infinitely superior to the widespread use of legal abortion, as in Japan today and more recently in some states of the U. S. We might even go so far as to predict that by the year 2000 many countries will have reached such a population density in relation to their food supply that no further increase can be tolerated. A marriage certificate might then bear two coupons entitling the couple to produce two children, no more. Restrictive tax measures, such as an income tax graduated more heavily as the number of children increases, or even temporary sterilization by court order, might be imposed by countries under desperation. Temporary sterilization of the female by implantation of a progesterone capsule would be effective enough.

Medical Aspects of Eugenics

We must now consider questions of eugenics, for complete and perfect control by individuals and by society over reproduction opens up certain eugenic possibilities like those of *Brave New World* (Huxley, 1932). First we must recognize that under our present application of ethics to medical practice, human society is already doing itself a considerable eugenic injury. When an infant or child with a genetic defect is kept alive by medical means and its defect is controlled or even eliminated, the child grows to adulthood. The most common sequel is marriage and parenthood. In the past, under a more cruel rule of nature, such persons never lived to reproduce, and their defective genes were thus eliminated from the population. The population thus maintained a balance between the defective genes being eliminated in each generation and the new ones being produced through mutation. (Not all mutations produce defective genes, but a considerable majority do so, for the simple reason that genes are the products of millions of years of selection of the best alternative means of guiding development in the normal environment and of working harmoniously together.) Medical practices thus tend to increase the frequency of defective genes in the population. To a certain extent, doctors are only making more work for themselves: since they have not removed the cause of the defect, which is the gene, but have only corrected symptoms, the defective gene when transmitted does the same harm once again.

116

Geneticists are looking forward to the day when they can practice genetic surgery or genetic manipulation; that is, really reach in and transform a defective gene and make it functional again. That will not be easy. The technique depends first of all on finding an effective but harmless human virus capable of transducing human genes from a donor of a sound gene to a recipient possessing a defective allele. The transduction would have to be done either in a very young embryo, before the cells have begun to multiply, or in the reproductive cells that actually produce a new individual. Before the technique is usable, then, we must perfect the means of culturing embryos and reproductive cells in the laboratory. Even so, it will probably always remain easier and simpler to discard defective reproductive cells and to select others from banked or cultured material that is free of known defect.

Now we come to the most difficult aspect of the crisis of values and goals. How can one select good strains of reproductive cells? If the same material is used to produce a great many embryos that are reared into babies, they will be too much alike—like members of the same caste in *Brave New World*. This difficulty might be avoided by never using a single line of reproductive cells more than a few times. There is another difficulty. Nearly all of us carry some defective genes; the average is probably around eight. We lack visible signs of defect because most defective genes are recessive; that is, they must be inherited in a double dose, coming from both father and mother, in order to produce an evident defect. As long as we have one working gene belonging to any particular pair of genes, enough of the protein it controls is produced to satisfy general needs. Close relatives, however, have a greatly heightened probability of carrying the same defective gene, because of possessing a common ancestor. It would therefore be necessary to have strict rules to prevent offspring being produced by persons who were derived from the same lines of banked or cultured reproductive cells, and careful records on the lineage of each person would have to be kept.

Detection of Harmful Genes

One new development seems certain to come

within a few years, since rapid strides are already being taken to make it possible. Tests are being developed that enable the laboratory specialist to determine whether or not a particular person carries even a single dose of a particular defective gene. For example, take the case of the disorder phenylketonuria (PKU), which in an untreated affected baby produces a certain kind of idiocy. The disorder results because the enzyme that transforms one amino acid into another is lacking. If you put such a child, very young, on a special diet with only a very little of the amino acid, phenylalanine, which is the one that cannot be properly utilized, the child will develop a normal mentality, and after it passes a certain age it will not relapse if taken off the special diet. Yet of course this individual has a double dose of the defective recessive gene and in adulthood will transmit the gene to its own offspring; and if the marriage partner also carries the gene, PKU may result. Prediction therefore depends on identifying the parents as carriers. It has been found that these parents are themselves not quite normal in their ability to utilize phenylalanine. They possess enough enzyme for ordinary purposes, and their mentality is in no way affected, but if injected with a rather large dose of phenylalanine they take about twice as long to get rid of it as do normal persons who lack the gene.

Thus we can detect carriers of the defective gene in the population, and advise them not to marry, or at least not to have children. In the past decade the ability to detect the carriers of specific recessive harmful genes has been extended to about 60 conditions. It consequently seems very likely to me that considerably before the year 2000 we will have genetic clinics in which by a battery of tests each prospective couple can learn whether or not both of them are carriers of the same defective genes. Advice to avoid having children and to substitute prenatal adoption could then be given. On the other hand, since the risk of having a defective child because of a specific recessive gene is only one in four when both parents are carriers, some couples may wish to take the risk. Whether advice or compulsion is to be used by society in these cases would seem to rest with the severity of the condition. If the prospective defect is one that would leave a baby a hopeless imbecile or idiot throughout life and a ward on society, or cause it to be born without limbs, or make it otherwise gravely defective, avoidance of parenthood ought to be mandatory. If, on the other hand, the condition is one that the new surgery or other methods of treatment might correct, as in the case of phenylketonuria, the risk might be taken. In any case, a new type of medical man, the human geneticist, is going to take his place along with the other specialists in the near future.

Questions of Freedom and Constraint

In a population suffering severely from overcrowding and subject to rigorous limitations of births, eugenics might be related rather simply to the measures for population control. For example, if a couple that had used up its coupons for two babies wanted additional children, they might be required to meet certain genetic tests before receiving a special permit. Some additional children above two per couple would be needed in some families to maintain the population at a given level, since some women have no children or only one, for a variety of reasons. The simplest eugenic test, yet one that in the long run might be quite effective in improving the population, would be simply to examine the first two children in order to assure that neither one was physically or mentally below average. Beyond the application of so simple a test, eugenic selection runs into frightful dilemmas. Who really possesses a "good" genotype? How do you judge, when what is the optimum genotype in one set of circumstances may be inferior in another? If we knew how to define the goal of a "good race" objectively, we might breed for it as we do animals; but the warning seems clear. In selecting for certain characteristics in their animal breeds, the breeders seem always to have sacrificed other, very desirable traits. The human races are not animal breeds, but each has been tested out by selection in a natural environment. Probably each is somewhat superior in its own way.

The control of human behavior by artificial means will have become by the year 2000 a frightening possibility. Government—"Big Brother"—might use tranquilizers, or hallucinogens like LSD, to keep the population from becoming unruly or overindependent. More and more subtle forms of conditioning will lead people to react in predictable ways desired by government or by commercial interests, without people quite knowing how they are hoodwinked. The added possibilities of controlled reproduction I have already described make these psychological methods of control over learning and behavior even more drastic. Here is our "brave new world" in full, with bottled babies in different kinds of solutions that would condition their mental growth to suit a certain caste. One wonders, moreover, what might be the effect of the complete liberation of the sexual life from its relationship to reproduction upon society and upon the family in particular. Recently Robert Morison (1967) of Cornell University, has pointed out the grave threat to the continuance of the family as the basic social unit. Can we safely, after a million years of human and prehuman evolution during which the family has been the basis of all protection, education, and nurture, give it up? What will be the psychological consequences of a population with no personal ties either to the older generation or to the younger generation? Can we look forward to the brotherhood of mankind when there are no more parents, brothers, or children—only people?

I have asked many questions that cannot at present be answered. I have predicted a future in which many cherished values of our society and many ethical standards may be questioned or superseded. It is not sufficient to have a few scientists raise such issues. Only prolonged and profound attention by many of the wisest men of our time, men of philosophy and religion, students of society and of government, and representatives of the common interests of men throughout the world, together with school administrators and scientists, may achieve a wise and sober solution of the crisis evoked in our world by scientific discoveries and their applications.

When is there overpopulation? Obviously, the question cannot be answered in simple terms. There is one answer to be given if you are thinking in the purely quantitative terms of numbers of people per square mile. But what sort of land is it? And what is the standard of living of the population? Quite another answer must be given if you are concerned with the quality of the population in genetic and evolutionary terms, and not merely with the state of its affluence. I am personally ready to accept the thoughtful definition I recently received from John A. Moore in an unpublished manuscript:

> *Overpopulation* exists in a unit of society when that unit uses the earth's resources at a greater rate than they are restored by natural or artificial means.

In that case Americans live in the most overpopulated land in the world, for they use and destroy, waste and dissipate the irreplaceable resources of the world at a rate that is estimated to be 50 times as great per person as it is for an Indian or African. Our enormous technological power is eating away the earth's limited supplies of fossil fuels and scarce metals at a rate that will exhaust them in a century, more or less. We devote little effort to the development of a new technology that will enable us to recycle and reuse materials to a maximum extent.

We are now entering a new era in the dimensions of human power—a day when we can modify the quality of the human population genetically as well as through improvement of the environment, a day when the course of human evolution will be placed in human hands. If we cannot solve even the relatively simple problems of controlling the quantity of human life on this so tiny planet, can we reasonably expect to manifest a greater wisdom in remodeling human nature and enlarging human capacities? We may foretell what man can be, but what *should* man be?

REFERENCES

Calhoun, J. B. 1962. Population density and social pathology. *Scientific American* 206 (2): 139-148.
Davis, K. 1963. Population. *Scientific American* 209 (9): 62-71.
Dobzhansky, T. 1962. *Mankind evolving.* Yale University Press, New Haven, Conn.
Glass, B. 1967. What man can be. *Educational Record* 48: 101-109. Separately printed by Thomas Alva Edison Foundation, Detroit.
_____. 1970. *The timely and the timeless.* Basic Books, New York. P. 15.
Hauser, P. 1960. Demographic dimensions of world politics. *Science* 131: 1641-1647.

HUXLEY, A. 1932. *Brave new world*. Reprinted 1958, Bantam Books, Harper & Bros., New York.

MALTHUS, T. R. 1798. *An essay on the principle of population*. Reprinted 1959 as *Population: the first essay*, Ann Arbor Paperbacks, University of Michigan Press, Ann Arbor.

MORISON, R. S. 1967. Where is biology taking us? *Science* 155: 429-433.

STRECKER, R. L. 1955. Populations of house mice. *Scientific American* 193 (12): 92-100.

———— and J. T. EMLEN. 1953. Regulatory mechanisms in house mouse populations: the effect of limited food supply on a confined population. *Ecology* 34: 375-385.

WRIGHT, S. 1960. Physiological genetics, ecology of populations, and natural selection. In *Evolution after Darwin*, ed. by S. Tax. University of Chicago Press, Vol. I, p. 429-475.

BIOETHICS: WHO DECIDES WHAT

INVIT: THE VIEW FROM THE GLASS OVIDUCT

The name has a vaguely Scandinavian sound with the accompanying promises of advanced sex. It is *Invit*, and it will be the result of the most advanced sex the planet has seen to date, though possibly not of the type you are thinking about.

Invit has been conceived not in a woman's body but in a laboratory apparatus. When he or she is born—and elements of the scientific community feel the birth will be within the next twelve months—Invit will be recorded as history's first known "ectoconceptus" (one conceived outside the womb), the progenitor of the test-tube baby.

Invit will be British. Disregard that misleading Scandinavian echo—the name is derived from the Latin *in vitro*, meaning a biological reaction taking place in an artificial apparatus, rather than within a living organism, *in vivo*.

The midwives are a group of intense and tight-lipped scientists and physicians led by Robert G. Edwards of Cambridge University's physiology laboratory. The actual birthplace, however, will be the Oldham General Hospital near Manchester, and the obstetrician-in-attendance is to be the hospital's Dr. Patrick Steptoe.

In the amphitheater (in thought at least) will be the world's most distinguished physiologists, including a vocal and highly critical group of Nobel laureates who believe that Edwards is committing an abominable act. James Watson of DNA fame, for example, has publicly demanded that Edwards abandon the Invit experiments. Max Perutz, also a laureate and a senior scientist for the British government's Cambridge-based Medical Research Council, has called Edwards's work a "stunt" and believes that "the whole nation should decide whether or not these experiments should continue." Common themes in both men's words are that Invit might be terribly deformed *à la thalidomide*, **resulting in a massive public backlash against all science, and that the Ed**wards technique—if successful—might open the door to a *Brave New World* form of genetic engineering.

Edwards's administrative superiors in the physiology laboratory, however, defend the work. The Edwards artificial conception approach has been tested exhaustively in mammals, they say, and occurs in a stage of embryonic development when the danger of there being birth defects is at its lowest.

Interestingly, much of the impetus behind the Invit experiments stems from an unlikely source—the prospective mother.

Mother?

She will be one of some fifty willing women chosen by Edwards for the creation of Invit. The women are principally in their mid-thirties. Some are doctors; others are doctor's wives or nurses. They are sterile, principally because of blockages in their oviducts. Their ova, consequently, cannot make contact with sperm cells. They and their husbands above all want to become parents, and so they turned to Edwards when his experiments became public knowledge more than a year ago.

Edwards's scientific feat was to find exactly the right hormonal moment to remove eggs from a female volunteer (by laparoscopy, a technique that involves a needlelike instrument inserted through the navel into the egg sac), the right way to select sperm donated by the husband, and the correct liquid medium to encourage both sperm and egg to interact. He succeeded not only in achieving fertilization of an egg by such means but in coaxing the embryo to divide more than 100 times. This is more than enough to prepare Invit for "implantation," or attachment to the uterine wall—the moment many scientists believe to be the time when life really begins.

Edwards plans to induce the creation of Invit by bringing about contact of egg and sperm in an ordinary cell-culture dish, putting the budding mixture into an ordinary laboratory incubator, and then implanting the embryo at the appropriate stage of division—into the woman's receptive uterus, again by laparoscope. The mother-to-be will then go through the usual nine months of waiting as would any pregnant woman.

Invit, then, will be born. If normal, Invit will be examined, studied, interviewed, coddled, and analyzed for the rest of its life—but discreetly, in the Cambridge way. Invit will be shielded from a curious world, perhaps not even told of his (or her) origin.

And nature will have been bypassed in her most intimate and awesome of acts, conception.

BABIES BY MEANS OF IN VITRO FERTILIZATION: UNETHICAL EXPERIMENTS ON THE UNBORN?

Leon R. Kass, M.D., Ph.D.

S INCE the disclosure of the unspeakable experiments performed by the Nazi physicians and
scientists upon helpless human beings, there has
been a growing interest in the difficult and important questions concerning the use of human subjects
in experimentation.[1-8] The principle of informed
consent has been enshrined as the ideal toward
which many practical steps are being and still need
to be taken. By making an effort to obtain consent,
the physician investigator is restrained from using
human beings simply as means to his own ends; by
giving his consent, the subject becomes a co-
adventurer with the physician investigator in the
search for new knowledge or new remedies for disease. But although there is agreement on what
ought to be, many difficulties remain in achieving
informed consent in practice.

Experimentation on children poses special problems. Because a child cannot give a truly informed
and voluntary consent, some have argued that all
nontherapeutic experimentation (i.e., research from
which the subjects cannot hope to derive therapeutic benefit) is unethical. Others, although con-
ceding the force of this objection, argue that the
physician has a moral duty to seek new means for
healing sick children — means that can only be
obtained by experimenting on them. Thus, in the
matter of experimentation on children, disagreement
on what ought to be done complicates the difficult
task of assuring ethical practices.

Given these theoretical and practical problems
in achieving ethical experimentation on existing
human beings, it is not surprising that little attention has been paid to questions concerning experi-

THE NEW ENGLAND JOURNAL OF MEDICINE, 1971, Vol. 285, pp. 1174-1179.

ments upon the unborn. Yet coming developments, including genetic manipulation of embryos or even gametes, surely will raise large ethical questions. More immediately, we face similar questions in connection with recently acquired abilities to initiate human life in vitro. This paper is an attempt to generate discussion of the ethics of experimenting on the unborn and the unconceived.

In Vitro Fertilization in Humans — State of the Art

Several laboratories have recently reported the in vitro fertilization of human egg by human sperm, and the subsequent laboratory culture of the zygote up to the blastocyst stage.[9-13] To surmount the difficulty of recovering eggs after their release from the ovary, Drs. P. C. Steptoe and R. G. Edwards have devised a surgical method, involving laparoscopy, to remove matured eggs directly from their follicles before ovulation.[14] From one woman, as many as three or four preovulatory oocytes can be recovered. Upon addition of sperm, fertilization occurs with a small but appreciable fraction of these eggs. Kept in culture medium, a majority of the fertilized eggs begin to divide, and a small fraction reaches the blastocyst stage, the stage at which the early embryo normally implants itself in the endometrium. Successful implantation of laboratory-grown blastocysts has been reported in rabbits and in mice, but not in humans.[15-17] The results in mice can be considered to be somewhat encouraging. A recent article reports that nearly half the transferred blastocysts developed into full-term, apparently normal progeny; however, the yield over all stages was low, with only 4 per cent of the starting number of eggs giving rise at the end to viable mice.[17] No gross abnormalities have been noted in any of the animals born alive after blastocyst transfer. And although some researchers would prefer to learn more about the control of implantation in animals before proceding further in the human work, others are inclined to go ahead in humans on a trial-and-error basis.

Ethics of Experimenting on the Unborn and the Unconceived

My discussion of ethical issues will consider only attempts to produce a child for a childless couple.

Thus, I am concerned here with the proposed implantation experiments, and not with laboratory fertilization itself. I shall examine some limited questions concerning the ethics of experimenting on human subjects, questions that will also apply to all proposed uses of this and future technologies to start new human lives. Other, broader questions in need of discussion lie beyond the scope of this paper; I have explored some of them elsewhere.[18]

The use of in vitro fertilization to initiate a new human life involves the necessary and deliberate manipulation of a human embryo, conceived and nurtured, at least for a time, in an artificial environment. Serious questions can be raised about the safety of the manipulations and of the environment and, hence, about the "normality" of any child whose conception and early development were subject to such manipulation. These medical questions about safety and "normality" lead to a perplexing moral question, since the hazards are being imposed on another human being, the prospective child, who obviously cannot consent to have such risks imposed upon him. The moral question is this: Does the parents' desire for a child (or the obstetrician's desire to help them) entitle them to have it by methods that deliberately impose upon that child an unknown and untested risk of deformity or malformation?

How unknown are the risks? Some of the leading researchers appear to disagree. Drs. Edwards and Steptoe are reported ready to proceed with implantation if tests on the embryos can rule out the presence of genetic or other defects (according to an article by Walter Sullivan in the New York Times, for October 29, 1970). Apparently, they are both concerned about the risks and ignorant of their likelihood. In their judgment, "[t]he normality of embryonic development and efficiency of embryo transfer cannot yet be assessed."[10] In contrast, Dr. Landrum Shettles does not talk about the need for further tests, and appears ready to proceed. In a recent article[12] he states that "the grossly normal blastocyst" was not transferred to the patient for the single reason that she had recently undergone uterine surgery. He adds: "Otherwise, there was no discernible contraindication for a successful transfer in vitro [sic — he means, in utero] and continued development. This is scheduled for patients with ligated or excised fallopian tubes who may want a child, with the ova obtained by culdoscopy or laparoscopy."*

The truth is that the risks are very much unknown. Although there have been no reports of gross deformities at birth after successful transfer in mice and in rabbits, the question of abnormalities has not been systematically investigated. No attempts have been made to detect defects that might appear at later times or lesser abnormalities apparent even at birth. In species more closely related to humans — e.g., in primates — successful in vitro fertilization has yet to be accomplished. The ability to produce normal young regularly by this method in monkeys seems to be a minimum prerequisite for use of the procedure in humans. The medical profession as a whole — and especially, professional societies of obstetricians and gynecologists — should press to see that this minimal requirement is met by all researchers.

But, even after normal young are produced in monkeys, we could not be certain that normal young would be produced in humans. There might be species differences in sensitivity to the physical manipulations or to possible teratogenic agents in a culture medium. Also, monkey experiments could neither rule out nor establish the risk of mental retardation for children resulting from experiments in humans. Unfortunately, as is often true, only humans can provide the test system for fully assessing the risks of using the procedure in humans.

Laboratory testing of the human embryos themselves, before transfer, cannot provide enough information about "normality," and might itself do damage. Gross abnormalities may be disclosed by ordinary microscopical observation. But ordinary microscopical observation can provide at best only a very crude measure of normality. Most genetic tests cannot be done on a given embryo without damaging it; moreover, there are few genetic tests presently

*The tone of the recent scientific articles and the reports of this research in the popular press suggest to the ordinary reader that a race may be on to the first embryo transfer. If such a race is on, it is likely that the swift will throw caution to the more sober; and will trust to luck that their victory in the race does not issue in a deformed or retarded child.

available for doing. Furthermore, damages might be introduced during the transfer procedure, even after the last inspection is completed. Conceivably, the manipulations might even make possible the implantation of some abnormal embryos that would have been spontaneously aborted if they had been generated under natural conditions.

In sum, there is at present no way of finding out in advance whether or not the viable progeny of the procedures of in vitro fertilization, culture and transfer of human embryos will be deformed, sterile or retarded. Even if we wished to practice abortion on all the misbegotten fetuses, we are not and will not be able to identify (by amniocentesis or other methods) many if not most of them. Neither can we count on "nature" to abort all of them for us.

The problem of risks and mishaps that accompany the experimental phase of this new technology provides a powerful moral objection sufficient to rebut the proposed implantation experiments. This moral objection should be widely shared, for it rests upon that minimal principle of medical practice, do no harm. In these prospective experiments upon the unconceived and the unborn, it is not enough not to know of any grave defects; one needs to know that there will be no such defects — or at least no more than there are without the procedure. The general presumption of ignorance is caution. When the subject-at-risk cannot give consent, the presumption should be abstention.*

It may be objected that all new medical technologies are risky, and that the kind of ethical scrupulosity I advocate would put a halt to medical progress. But such an objection would ignore a crucial distinction. It is one thing to accept for yourself the risk of a dangerous procedure (or to consent on behalf of your child, even your intrauterine child) if the purpose is therapeutic. Some might say that this is not only permissible, but obligatory, in line with a duty to preserve one's own health or the health of one's children. It is quite a different thing deliberately to submit a child, born or unborn, to hazardous procedures which can in no way be considered thera-

*I would apply a similar argument to the case in which husband and wife are known to be carriers for the same severe, recessive, genetic disease. Such would-be parents would act responsibly by abstaining from procreation, and by adopting children instead. Whether the state and the law ought to compel this abstention is a wholly different question.

peutic for him (and, as I shall argue shortly, is "therapeutic" for you only in that it "treats" your desires, albeit unobjectionable ones). This argument against nontherapeutic experimentation on children applies with even greater force against experimentation "on" a hypothetical child (whose conception is as yet only intellectual). One cannot ethically choose for him the unknown hazards that he must face and simultaneously choose to give him life in which to face them. This judgment could be set aside only under a strongly pronatalist ethic — much more pronatalist than Roman Catholicism ever was — which would hold that parents have either a pre-eminent duty or a pre-eminent right to have their own biologic child by whatever means.

EXPERIMENT OR THERAPY?

At attempt to produce a live baby by means of laboratory fertilization and culture certainly must be deemed "experimental" in the sense of being new and untested, full of uncertainties and unknown hazards. Yet the use of untested, potentially hazardous procedures can often be justified if the purpose is therapeutic (i.e., "non-experimental"), and if the likely therapeutic benefits outweigh the likely risks. But is the purpose here therapeutic, or might it not also be labeled as "experimental" (i.e., nontherapeutic, or scientific) in this second sense as well? Clearly, the procedure is not therapeutic for the child generated. However, the first attempt to produce a live baby with in vitro fertilization will most probably be described as serving a therapeutic purpose for the parents — namely, the treatment of their infertility. But is this an accurate description? Is the inability to conceive a disease, or merely a symptom of disease? Can a couple have a disease? Does infertility demand treatment wherever found (e.g., in women over 70 or in virgin girls?) or by any and all available means (e.g., by artificial insemination, by in vitro fertilization, by extracorporeal gestation, by parthenogenesis or by asexual reproduction or cloning)?

Infertility is not a disease in the usual sense, but it can be a symptom of disease. It is not life threatening or crippling, nor does it lead to detectable bodily damage. To consider it a disease leads to a focus on an individual; yet infertility is a condition that is located in a marriage, in a union of two individuals. If it is any kind of disease, it is a "social disease." Even though the abnormality responsible is usually found in only one of the partners, their interaction is required to make the problem manifest.

More is at stake here than the correction of linguistic imprecision; the error in language is not innocuous. To consider infertility solely in terms of the traditional medical model of disease (or in terms of a so-called right of an individual to have a child) can only help to undermine, both in thought and in practice, the bond between childbearing and the covenant of marriage. In a technological age, viewing infertility as a disease demanding treatment by physicians fosters the development and encourages the use of all the new technologies mentioned above.

Just as infertility is not a disease, so providing a child by artificial means to a woman with blocked oviducts is not treatment (as surgical reconstruction of her oviducts would be). She remains as infertile as before. What is being "treated" is her desire — a perfectly normal and unobjectionable desire — to bear a child. There is no clear medical therapeutic purpose that requires or demands the use of the new and untested technologies for initiating human life and that might possibly justify the unconsented-to use of a human subject for the benefit of others and at risk to himself.

Is There a Human Subject?

My use of the term "human subject," as applied to a prospective child, requires some justification. Who is the human subject of the fertilization, culture and implantation procedures? Do we mean to call a blastocyst a human subject? The questions as raised here can easily be distinguished from a similar question raised concerning abortion. With abortion, the moral question is when and whether one can justify killing the fetus; one of the underlying issues is whether the fetus is a human being or a potential human being worthy of protection. Here, we are concerned with the possible harm inflicted upon the live, breathing children who come to be born after getting their start via in vitro fertilization and laboratory culture. The underlying issue is whether one can speak of such children and their ontogenetic precursors as "human subjects of experiments," especially when they are themselves the products of such experiments.

The issue can be clarified and possibly resolved if analogous harmful manipulations of the unborn are considered. A useful, though not perfect, analogy would be the deliberate administration of thalidomide or some other known teratogen to a pregnant woman. Whether the fetus is then "human" or not, the child it becomes would be an unwilling and unjustly injured victim of such unethical practice. More analogous would be the generation of a child through artificial insemination with sperm that had been deliberately irradiated or mutagenized. These examples, although hypothetical, serve to show that a child can be deliberately injured before his birth, even coincidentally with his conception. If we ask who it is that was injured when the injury occurred, the answer must be, "the human subject," or at least, the "potential human subject," who is the prospective child. And if the child-to-be can be deliberately injured, he can be negligently injured, as he might be if he were the product of in vitro fertilization before the technic were shown to be free of risk.

This line of reasoning has already found its way into the law, and is likely to gain strength as the technologies for manipulating the unborn proliferate, and as we learn more about specific harmful effects of drugs and chemicals on the fetus. Indeed, legal actions claiming intentional and negligent wrongdoing have recently been brought by infant plaintiffs, against their parents or against doctors and hospitals, seeking damages for "wrongful life" — i.e., life with inextricably linked handicaps such as illegitimacy or congenital syphilis.[19] Although judges face grave policy questions in awarding damages to the plaintiff in these cases (none have done so as yet), several of their opinions affirm that children can be and have been injured while unborn, and even at the moment of conception.

EXPERIMENTS WITH CHILDLESS COUPLES

Most of the scientific reports on human-embryo experimentation are strangely silent on the nature of the egg donors, on their understanding of what was to be done with their eggs, and on the manner of obtaining their consent. This silence is surprising in view of the growing sensitivity of the medical and scientific communities to the requirement of informed consent, and especially surprising given the kind of experiments here being performed. In a recent article in Scientific American by Edwards and Fowler,[20] there is this solitary comment: "Our patients were childless couples who hoped our research might enable them to have children." From the report that they had hopes, we can surmise that they considered themselves to be patients. But so far as these experiments are concerned, they are only experimental subjects. The researchers owe us an account of how consent was obtained and of what the couples were in fact told.

Only one of the scientific articles,[14] that describing the use of laparoscopic surgery to recover human oocytes, tells us anything more about the persons used as experimental subjects, and about how they were informed: "The object of the investigations was fully discussed with the patients, including the possible clinical applications to relieve their infertility." Though welcome, this statement is incomplete. It does not tell us whether the couples were also informed that the much more likely possibility was that it would be future infertile women, rather than they themselves, whose infertility might be "relieved." At this stage of technical competence, the likelihood of failure must be made clear when consent to undergo laparoscopy is obtained.

It is altogether too easy to exploit, even unwittingly, the desires of a childless couple. It would be cruel to generate for them false hopes (e.g., by exaggerated publicity). It would be both cruel and unethical to generate hope falsely (e.g., by telling women that they themselves, rather than future infertile women, might be helped to have a child) to obtain their participation in experiments.*

---

*That this may have already occurred is suggested by the following extract from a news report by Patrick Massey, of Reuters, that appeared in the Washington Post, March 3, 1970: "Dr. Patrick Steptoe, who heads the team of doctors working on the experiment, disclosed on television that he had extracted an ovum from a 34-year-old housewife and fertilized it with her husband's sperm. The woman, Mrs. Sylvia Allen, . . . said she hoped the fertilized ovum would be implanted in her womb in the next two to six weeks, meaning that the world's first baby conceived in a test-tube could be born by the end of 1970." The implantation was never performed.

The arguments that I have developed, in this paper and elsewhere,[18,21] have led me to the conclusion that we need regulation of the technologic application of research in human reproduction (though not necessarily of the research itself). Some of the regulation can and should be done by scientists and physicians, and by their professional societies. Such intraprofessional self-scrutiny has been responsible for much of the progress made in recent years in improving the standards for human experimentation, for adequate drug testing and for care of laboratory animals. Regarding the particular issue of human experimentation raised in this paper, I suggest to my colleagues the following specific steps:

The first would be a profession-wide, self-imposed moratorium on attempts to produce new human children by means of in vitro fertilization and embryo transfer (and by other new procedures), *at least* until such time as the safety of the procedures can be assessed and assured.

The second would be initiation of critical, prospective studies in primates and other mammals to assess the "normality" of the young produced by artificial means.

The third would be establishment of intraprofessional bodies and forums to discuss and to evaluate critically work in mammalian and especially in human reproduction. Reports by such responsible professional groups could help to prevent the creation of inflated hopes and fears.

Yet the questions of risk to the unborn and of informed consent are only a few questions among many that must be raised. Would the assurance of safety and normality provide a sufficient warrant for going ahead? Surely, there is more at issue than providing a child for an infertile woman. Once introduced for that purpose, laboratory fertilization can be used for any purpose. Indeed, the work described is a giant step toward the full laboratory control of human reproduction. What are the implications of this step, and of the others it makes possible (such as ectogenesis or cloning[22]) — e.g., for the human-ness of human procreation or for the human family? Should not the weighing of ethical and social considerations concerning both the widespread use and subsequent uses of the new technology enter into the decision to apply it for the first time?

These ethical questions point to a political question. Laboratory control of fertilization and embryonic development is a major departure in human procreation whose human consequences, both private and public, are likely to be profound. It is no mere ordinary medical (i.e., therapeutic) advance. Therefore, one must question the wisdom of leaving the decision to go ahead for the private judgment of a team of physicians and scientists (whose judgment I am not now questioning), or even for the collective judgment of the medical and scientific community. Is this not a matter that deserves broader public deliberation and, in the end, might be one for public decision?

In the light of these remarks, I make the following additions to my list of specific proposals for physicians and scientists, and their professional societies:

Initiation of inter-disciplinary discussion, both in and out of the government, of the desirability of introducing the new technologies, and of the means for anticipating and minimizing the undesirable social consequences, if they are introduced.

Co-operation with lawyers, legislators, theologians, philosophers, humanists, social scientists and laymen in establishing ethical guidelines for the use of reproduction technology, and in providing for the proper legal safeguards for experimental subjects, including unborn children.

Convocation of international groups to consider desirable, necessary and feasible means of preventing follies and evils committed in the name of international competition.

Some, if not all, of these suggestions are likely to be unpopular, and some of them are not without their own dangers. But they should not be seen as harbingers of the bogeyman who goes around stopping research, but rather as suggestions for constructive and responsible steps that our growing power over human life obliges us to take. It should not be forgotten that our society already exercises considerable control over the rate of technologic development in this area, by means of its granting power and by the power that it confers on scientists and physicians. Thus, the question is not control versus no control, but rather what kind of control, by whom, and to what purpose. If doctors and scientists continue to be shortsighted, if they fail to regulate themselves and to co-operate with other professions and groups with legitimate concerns in these areas, they can only expect society to react with more sweeping, immoderate and throttling controls in the future. One need only consider the likely public reaction if the first "test-tube-baby" turns out to be a monster.

Yet this concern for scientific self-preservation, for the right to experiment, is insufficient. With our growing power to affect the lives of unborn children, through new technics for beginning life and coming technics for genetic manipulation, we scientists and physicians have a growing responsibility to that broader community to which we belong, the human race, and especially, to each human being upon whom we exercise our power.

REFERENCES

1. Beecher HK: Ethics and clinical research. N Engl J Med 274: 1354-1360, 1966
2. Clinical Investigation in Medicine: Legal, ethical, and moral aspects: An anthology and bibliography. Edited by I Ladimer, RW Newman. Boston, Law-Medicine Research Institute, Boston University, 1963
3. Pappworth MH: Human Guinea Pigs: Experimentation on man. Boston, Beacon Press, 1968
4. Fletcher J: Human experimentation: ethics in the consent situation. Law Contemp Probl 32:620-649, 1967
5. Ethical Aspects of Experimentation with Human Subjects. Daedalus 98:219-594, 1969
6. New Dimensions in Legal and Ethical Concepts for Human Research. Ann NY Acad Sci 169:293-593, 1970
7. Beecher HK: Research and the Individual: Human studies. Boston, Little, Brown and Company, 1970
8. Ramsey P: The Patient as Person: Explorations in medical ethics. New Haven, Yale University Press, 1970
9. Edwards RG, Bavister BD, Steptoe PC: Early stages of fertilization *in vitro* of human oocytes matured *in vitro*. Nature (Lond) 221:632-635, 1969
10. Edwards RG, Steptoe PC, Purdy JM: Fertilization and cleavage *in vitro* of preovulator human oocytes. Nature (Lond) 227:1307-1309, 1970
11. Steptoe PC, Edwards RG, Purdy JM: Human blastocysts grown in culture. Nature (Lond) 229:132-133, 1971
12. Shettles LB: Human blastocysts grown *in vitro* in ovulation cervical mucus. Nature (Lond) 229:343, 1971
13. Jacobson CB, Sites JG, Arias-Bernal LF: *In vitro* maturation and fertilization of human follicular oocytes. Int J Fertil 15:103-114, 1970
14. Steptoe PC, Edwards RG: Laparoscopic recovery of preovulatory human oocytes after priming of ovaries with gonadotrophins. Lancet 1:683-689, 1970

15. Chang MC: Fertilization of rabbit ova *in vitro*. Nature (Lond) 184:466-467, 1959
16. Whittingham DG: Fertilization of mouse eggs *in vitro*. Nature (Lond) 220:592-593, 1968
17. Mukherjee AB, Cohen MM: Development of normal mice by *in vitro* fertilization. Nature (Lond) 228:472-473, 1970
18. Kass LR: New beginnings in life, Three Medical Futures. Edited by M Hamilton. Grand Rapids, Michigan, William B Eerdmans Publishing Company (in press)
19. Tedeschi G: On tort liability for "wrongful life." Isr Law Rev 1: 513-538, 1966
20. Edwards RG, Fowler RE: Human embryos in the laboratory. Sci Am 223 (6):44-54, 1970
21. Kass LR: Freedom, coercion, and asexual reproduction, Freedom, Coercion, and the Life Sciences. Edited by D Callahan, LR Kass. Cambridge, Harvard University Press (in press)
22. Watson JD: Potential consequences of experimentation with human eggs. Presented at the twelfth meeting of the panel on science and technology, Committee on Science and Astronautics, United States House of Representatives, Washington, DC, January 28, 1971

137

Moral and Legal Decisions in Reproductive and Genetic Engineering

WERNER G. HEIM

Man has long known that he can modify the genetic
bases of organisms by controlling their breeding
patterns. The application of such knowledge to the
development of specific breeds of dogs or of wheat
strains particularly suitable to cultivation is quite
old. Mendelian and post-Mendelian genetics increased this ability by providing a rational model for
the observed effects and thereby increased the ability
to predict. This ability, when combined with advanced knowledge in the physical sciences, led to an
understanding of many of the major mechanisms for
the genetic transmission of information. Finally, this
newest knowledge is rapidly providing the means for
the manipulation of that hereditary information itself.

Imminence of the Developments

The ability to change the hereditary information
content of a human cell is not a matter for the future:
it is presently available. In 1971 Merrill, Geier, and
Petricciani showed that a gene—a unit of hereditary
information—from a bacterium could not only be
introduced into human cells but could be caused to

THE AMERICAN BIOLOGY TEACHER, 1972, Vol. 34, pp. 315-318.

function in them. Nor need we any longer simply search for a gene that we may wish to manipulate: in 1972 a human gene was synthesized by Kacian and his associates. These techniques are still laboratory exercises; but the time-span from first demonstration to practical application is not likely to be longer than for similarly important developments in the physical sciences. Transistors, for instance, were in general use within less than a decade of their initial development.

An additional line of research is likely to accelerate the application of genetic manipulation, or genetic engineering, to man. This is what may be termed reproductive engineering. Reproductive engineering is any deliberate manipulation of the procreative part of the life cycle. Much of this is already in daily use in such forms as conception control and artificial insemination.

Genetic and reproductive engineering tend to be synergistic. Consider, for example, the production of genetically mosaic mice; that is, mice whose heredity is literally a patchwork. This condition is achieved by fusion of very early embryos of different strains. This requires at least four interacting techniques: (i) the exact timing of pregnancies, (ii) the manipulation of the eight-cell embryos in vitro, (iii) the production of exactly timed pseudopregnancies in the recipients, and (iv) the safe implantation of the mosaic, tetraparental embryos into the pseudopregnant females. The technique of transferring very early embryos from the uterus of one female to the uterus of another—this is called inovulation—itself has implications, to be discussed later. Another potential synergism between the two kinds of engineering may soon appear in the form of the insertion of genes into sperm prior to the use of that sperm in artificial insemination.

A warning should be added here, however. In all of these processes highly specific techniques must be used. This means that one kind of manipulation may be possible at a particular time while another, and apparently closely related, procedure remains out of reach. For example, it is now possible to insert a gene into a human cell; but to remove a specific gene is not yet possible, and indications are that it may remain impossible for a long time. It follows that only a person well acquainted with both the scientific literature and the conceptual basis of these activities is in any position to estimate the time scale for a future development.

Uniqueness of the Developments

These new scientific developments, which have been and are being incorporated into the technology of our society, pose serious moral and legal questions. That, of course, is not in itself new; scientific developments, when more or less broadly applied, have always raised such questions. (Recent examples include the fluoridation controversy and the legal aspects of artificial insemination.) What is new is the nature of the questions raised by the development of genetic and reproductive engineering: the manipulations of man, in these fields, will be vastly more fundamental than any previous ones. In some instances the manipulations will be irreversible, not only with respect to the individual but also with respect to all his descendants and to the population. In such circumstances wrong decisions can lead not only to the death of the individual but also to extinction of the species. Other incorrect decisions could seriously degrade the quality of life through control of thought, of liberty, and of motivation.

Roles in Controlling the Applications

Unfortunately the track record of those who, in my opinion, should be developing the moral positions, public policies, and legal constraints concerning these developments is not good. With few exceptions the moralists, theologians, sociologists, political scientists, legislators, legal scholars, and judges, whose job it is to integrate new developments into a sustained, viable, and sound fabric of society, have not only failed to be prepared for the introduction of new, science-derived technologies into the culture but have even reacted, at best, rather sluggishly. Their reaction time has sometimes been slower than that of the general public, as is shown by the widespread acceptance of certain developments long before they are placed in a definite legal and moral framework. An example is the debate about "when and whether to pull the plug" of the machine keeping a terminal patient more or less alive. This debate commenced seriously only well *after* the machines had come into common use and is still often being conducted in terms hardly useful to the person who must make the decision: the physician on the ward. Another example of the failure to develop an adequate set of moral stances has to do with genetic counseling and the individual and societal risks arising from the reproduction of persons who are sustained by medical skills in the face of hereditary defects. Should an 11th commandment have been promulgated: "Breed not, ye who carry defects"?

One popular reaction to this lack of guidance has been a feeling that research in these fields should be stopped because "it's too dangerous." Of course it is not the research itself that is dangerous but the application of the results within a society that has developed neither the broad principles nor the specific directives to make these applications wisely.

By default, the natural scientist himself has had to fill the vacuum with his own inexpert opinion so frequently that the impression is now abroad in some circles that this, too, is a part of his job. The job of the natural scientist is to make the discoveries; that of the technologist is to develop applications; and that of the social scientist and humanist is to suggest whether, how, and under what conditions the work of the other two ought to be applied.

As a well-informed citizen the natural scientist should have, of course, the same opportunity and at least the same responsibility to contribute to the decision-making process as any other well-informed citizen. But he is not an expert in, and should not be expected to act as if he were an expert in, the delicate processes of weighing the data or of drawing **conclusions and making recommendations. He does, however, have responsibilities in this constellation of events beyond those of the ordinary person. First, he should give plenty of advance notice to the humanists and social scientists of forthcoming developments likely to require their attention. In doing so he must put all his expertise to work to differentiate between science and science fiction, lest the world react like those to whom the shepherd cried "wolf" too often. Second, he must be prepared to give the relevant details of new developments to his colleagues in the humanities and the social sciences. He must do this with due regard for (i) relevancy,** lest he either swamp or impoverish the communication channels; (ii) scrupulous accuracy, lest he mislead his listeners; and (iii) intelligibility of language, lest he be misunderstood. Finally, he ought to monitor the tentative and the definitive pronouncements of the humanists and social scientists so as to detect early any misinterpretation or lack of facts that may have distorted their work.

The need for careful formulation of moral and legal positions on new developments *before* their widespread use is now more critical than ever before: the changes are more fundamental in nature, are less likely to be reversible in the individual or in his descendants, and, most importantly, are changes in human nature itself. As Leon Kass (1971) has pointed out, ". . . both those who welcome and those who fear the advent of 'human engineering' ground their hopes and fears in the same prospect: *that man can for the first time recreate himself*. Engineering the engineer seems to differ in kind from engineering his engine." (Kass's emphasis.)

Aspects of Human Engineering

The "human engineering" of which Kass speaks— it is inclusive of what is here called genetic and reproductive engineering—differs from previous kinds of changes in several respects. Consider its influence on human conduct. The traditional methods of modification have had three important characteristics: (i) they used symbols, especially as embodied in speech and art, as their primary vehicle; (ii) they allowed considerable choice to the individual as to the acceptance of at least parts of the modifications offered; and (iii) their effects could, to a great degree, be reversed in both the individual and his progeny. In contrast, the changes brought about by human engineering are largely nonsymbolic, because they modify the conduct-controlling mechanism directly. For example, the human engineer would *not* seek to educate or train a victim of Down's syndrome (the so-called mongolian idiot) to the point of self-sufficiency but rather would eliminate the chromosomal defect or, alternatively, would block fertilization or development of eggs carrying the defect. This is, incidentally, in sharp contrast with one of the most advanced of the old-style techniques, Skinnerian operant conditioning.

The second contrast between the old and the new techniques lies in the fact that the individual frequently will have no power to adopt or reject a particular modification. The modification will have occurred before he became an individual—either in the gametes or the early embryo that produced him or in an ancestral generation. To the extent that the change is in the hereditary material itself, it may be an irreversible change for either of two reasons: the technique for reversing it may not be available— at present a gene may be added but not subtracted— or else the desire to make the reversal at some future date may have been blocked genetically at the same time that the other changes were made.

A further consequence of the difference between the old and the new techniques is that the new techniques have the effect of removing literature, mythology, religious liturgy, etc., one more step from the real seat of power. Once, long ago, a speech—a curse—was expected to kill an enemy. Later a speech was expected to inspire soldiers to kill the enemy. In the future a speech might cause technicians to change the genes of some persons so that, under certain conditions or at a certain age, these persons simply die.

"Now" and "Soon" Examples

What, then, are some of the more likely of these human-engineering techniques? I wish here to deal only with those that are either available right now or have a great likelihood of becoming available in the next few years. Due to the highly technical nature of the work, long-range prophecy is likely to be unprofitable.

One technique that fits the human-engineering category, although it is neither reproductive nor genetic, is that of the direct control of man's neural system by electrical or drug stimulation of specific parts of the brain, as detailed so well by Delgado (1969). One need only read his book to discover the tremendous power available here and the need to develop a proper framework for that power.

Other nonreproductive and nongenetic methods are those devised for the prolongation of life by what, at any point in time, would be considered "extraordinary" means: heart transplantation, heart prosthesis, and the like.

In the realm of genetic engineering itself, one must at least mention genetic counseling—already widespread and rapidly spreading—because serious moral problems crop up almost daily in this work. If a prospective child has a risk of 5% that he will be seriously defective, should the parents reproduce? Or, if conception has occurred, should an abortion be done? What if the risk is 10% or 30% or 60%? What if the risk is essentially zero that the child will be affected but quite substantial that the child will transmit the deleterious gene?

Let us turn now to questions of genetic engineering in the strictest sense. Here we are dealing with techniques that change "human nature" and are transmissible directly to the offspring. Under what conditions should such techniques be used? The technique for inserting a gene or a group of genes into sperm is likely to be practical within a very few years. Suppose two persons are defective in the gene for the production of the enzyme parahydroxylase: their child may suffer from the disease phenylketonuria (PKU). It should be possible soon to introduce the normal, active gene into the sperm of the father and thereby assure that the cells of the offspring have, each, at least one properly functioning gene for this enzyme. Should this procedure be done? Consider not only that normal offspring will be produced but that both of the defective genes will be available to, although perhaps not active in, further progeny.

In the eugenic uses of genetic engineering that I have mentioned here, most of the arguments seem to be on one side: because it is highly likely that both the individual and society will benefit from the engineered changes, and because either short-term or long-term deleterious effects are unlikely, the judgment probably will be that the procedure should be carried out. But let's take a less simple case. There is good evidence for a hereditary component in the origin of schizophrenia and schizoid conditions (Heston, 1970). Very likely we are dealing with a multigenic inheritance pattern here. Therefore there may be a defective gene that facilitates the appearance of a mild schizoid condition. It is not unreasonable to suppose that it should become possible to insert a gene into the fertilized egg—a gene that would "correct" the defective gene. At first sight this would seem to be another clear-cut case for the use of the technique. But there are reasons to suppose that mild schizoid tendencies are useful in a wide range of endeavors, from painting and composing to scientific research. It is obvious that, without *much* further information, both the long-term and the short-term benefits cannot be weighed properly against the possible harm. And remember that once such a corrective gene has been introduced, it may not be removable! In such an equivocal situation what ought to be the rules for the application of the technique? I submit that, in such a situation, the role of the biologist is to provide the raw information and, perhaps, an indication of the probabilities of the various consequences. Then the humanists and social scientists should formulate the rules, subject to revision as more information or better techniques become available.

The other type of human engineering—reproductive engineering—also will require much thought and study. Although less likely to have permanent or irreversible effects, its points of action are even more closely bound up with our general ideas of morality and proper conduct than are those of genetic engineering. This is evident from the arguments raging around the two early forms of reproductive engineering presently in use: conception control and artificial insemination. Neither legal nor moral opinion has been stabilized or universalized in respect to these. What are and what should be the legal and moral positions vis-à-vis the following procedures, all but one of which have already been carried out in nonhuman organisms?

1. Artificial insemination with sperm from donors selected for particular qualities.

2. Artificial inovulation in which an egg produced by the wife and fertilized by the husband is transferred to a foster uterus.

3. Artificial inovulation by transfer of an egg produced by a donor, and fertilized either by the husband or a sperm donor, into the uterus of the wife.

4. Insertion of the nuclei of ordinary body-cells into eggs whose own nuclei have been removed, producing thereby any desired number of individuals with exactly the same genetic constitution as that of the donor of the nuclei—a procedure called cloning.

5. Cloning by causing ordinary body-cells to act as if they were fertilized eggs, thereby (again) producing any desired number of persons having the same hereditary makeup as the donor of the cells.

So little thinking has been done about the last two of these possibilities (outside the realm of science fiction, anyway) that we can hardly even list the individual or social advantages and disadvantages that might accrue from having a large number of genetically identical individuals. We are even less in a position to give these possibilities the calm and deliberate weighing that ought to lead to acceptable patterns of application.

Prospects and Challenges

The following general conclusions can be drawn:

1. Specific modification of individuals, by action on the genes or on the reproductive patterns and for individual or social purposes, is now possible and is increasingly and rapidly becoming feasible.

2. These techniques differ fundamentally from older approaches based on restricted breeding patterns—classical eugenics and Hitlerian eugenics, for example—in two critical respects: they are fast and they work. Even the insertion into man's heredity of what might be termed socially specific genes is a rather close—an uncomfortably close—probability.

3. We need to develop ways of adjusting these possibilities, of restricting their use within morally and socially acceptable patterns.

4. To do this will require a rejuvenation of the humanities and the social sciences and a reshaping of the relationship between them and the natural sciences.

Athelstan Spilhaus (1972) has said: "Just as technological invention cannot remove the need for social invention, neither should our slowness in changing outmoded social practices, institutions, and traditions be allowed to slow technological realization of potential benefits to all." Unless we rearrange our houses so that we can get to work immediately and unless we do get to work immediately, the genie (or gene) will be out of the bottle before we know the magic formulas that control it. And to permit this kind of genie to be out of control or to be misused is (to use the words in an unusual but meaningful combination) to *commit extinction*—first of freedom, and then of the species.

As teachers we have additional duties: to inform our students of these developments and to help them make the questions raised by these developments a prime order of business in their lives. We may confidently expect their lives to encompass the period during which most or all of the developments outlined here, as well as others, will be brought into practice. Those who are presently our students will be the ministers, the judges, the political scientists, the philosophers, the legislators, the teachers and, yes, the biologists, in charge. We and they had better start to think fast about how to handle that charge.

REFERENCES

Delgado, J. M. R. 1969. *Physical control of the mind: toward a psychocivilized society.* Harper & Row, New York.

Heston, L. L. 1970. The genetics of schizophrenic and schizoid disease. *Science* 167:. 249-256.

Kacian, D. L., *et al.* 1972. In vitro synthesis of DNA components of human genes for globins. *Nature New Biology* 235: 167-169.

Kass, L. R. 1971. The new biology: what price relieving man's estate? *Science* 174: 779-787.

Merrill, C. R., M. R. Geier, and J. C. Petricciani. 1971. Bacterial virus gene expression in human cells. *Nature* 233: 398-400.

Mintz, B. 1971. Genetic mosaicism in vivo: development and disease in allophenic mice. *Federation Proceedings, Federation of American Societies for Experimental Biology* 30: 935-943.

Spilhaus, A. 1972. Ecolibrium. *Science* 175: 711-715.